Atmosphere, Earth, Ocean & Space

The series Atmosphere, Earth, Ocean & Space (AEONS) publishes state-of-art studies spanning all areas of Earth and Space Sciences. It aims to provide the academic communities with new theories, observations, analytical and experimental methods, and technical advances in related fields. The series includes monographs, edited volumes, lecture notes and professional books with high quality. The key topics in AEONS include but are not limited to: Aeronomy and ionospheric physics, Atmospheric sciences, Biogeosciences, Cryosphere sciences, Geochemistry, Geodesy, Geomagnetism, Environmental informatics, Hydrological sciences, Magnetospheric physics, Mineral physics, Natural hazards, Nonlinear geophysics, Ocean sciences, Seismology, Solar-terrestrial sciences, Tectonics and Volcanology.

More information about this series at http://www.springer.com/series/16015

Ching-Hua Lo · Xiwei Xu · Wen-Yen Chang ·
Masataka Ando
Editors

Earthquake Geology and Tectonophysics around Eastern Tibet and Taiwan

Springer

Editors
Ching-Hua Lo
Department of Geosciences
National Taiwan University
Taipei, Taiwan

Xiwei Xu
China Earthquake Administration
Institute of Crustal Dynamics
Beijing, China

Wen-Yen Chang
Eastern Taiwan Earthquake
Research Center (E-TEC)
National Dong Hwa University
Hualian, Taiwan

Masataka Ando
Center for Integrated Research and
Education of Natural Hazards
Shizuoka University
Shizuoka, Japan

ISSN 2524-440X ISSN 2524-4418 (electronic)
Atmosphere, Earth, Ocean & Space
ISBN 978-981-15-6209-9 ISBN 978-981-15-6210-5 (eBook)
https://doi.org/10.1007/978-981-15-6210-5

This Springer imprint is published by the registered company Springer Nature Singapore Pte Ltd.
The registered company address is: 152 Beach Road, #21-01/04 Gateway East, Singapore 189721, Singapore

Preface

In the past 20 years, Asia's two largest earthquakes—the 1999 Chi-Chi earthquake and the 2008 Wenchuan earthquake—occurred at crustal deformation fronts along the most active collision zones in Asia. The Chi-Chi earthquake occurred within the Taiwan arc–continent collision zone, and the Wenchuan earthquake occurred within the India–Eurasia continent–continent collision zone. The impact of these earthquakes is reflected not only by devastating socio-economic losses, but also by the sheer number of people affected. Hundreds of scientific papers have subsequently been published to investigate these earthquakes, including Cross-Strait collaborations among scientists from Taiwan and mainland China. The present volume is a compilation of review articles that summarize our current understanding of the mechanisms, kinematics, crustal structure, and faulting responsible for the Chi-Chi and Wenchuan earthquakes.

This volume provides a holistic geological and geophysical background to understand how these earthquakes occurred. It starts with two articles (Huang and Chang; Xu and Xu) describing the co-seismic rupture and surface deformation caused by the 2008 Wenchuan earthquake, followed by Tan et al. discussing the long deformation history of the Longmenshan thrust belt where the Wenchuan earthquake is located. Ma's article summarizes the study results of the 1999 Taiwan Chi-Chi earthquake and the Taiwan Chelungpu-fault drilling project and discusses earthquake kinematics and dynamic processes from seismic waveform modeling. This is followed by two articles focused on earthquakes associated with active fault in the eastern margin of the Tibetan Plateau and the thrust belt in Taiwan: (1) Lu et al. discuss the seismotectonics of the Lushan earthquake, and (2) Zhao et al. discuss the stress environment in South Taiwan. Chang et al. review seismo-geological signatures for $M \geqq 7.0$ earthquakes and their experimental prediction in Taiwan. In the last contribution, Cai et al. give a broad picture of tectonic structures in the Cathaysia block in China and discuss its potentials for large earthquakes.

In summary, the articles encompass a broad range of earthquake topics, not only mechanisms and kinematics, but also the geological and geophysical background of the fracture faults involved. It is hoped that the volume can serve as a helpful reference book, not only for the two disastrous earthquakes mentioned above, but also for the study of similar tectonic settings where earthquakes are likely to occur.

Taipei, Taiwan — Ching-Hua Lo
Beijing, China — Xiwei Xu
Hualian, Taiwan — Wen-Yen Chang
Shizuoka, Japan — Masataka Ando

Contents

Chapter 1
Complicated Source Rupture Process and Induced Coseismic Surface Deformation of the Wenchuan M_W 7.9 Earthquake and Their Tectonic Implications

Bor-Shouh Huang and Chung-Pai Chang

Abstract The source rupture processes and coseismic surface rupture of the Wenchuan earthquake are recognized as having the complicated patterns of a great earthquake. In this article, we review some of the Wenchuan earthquake source studies conducted by scientists from Taiwan and mainland China under the scientific cooperation of both sides of the Taiwan Strait. To clarify how these ruptures developed and their tectonic implications, we report the source imaging using two regional seismic arrays based on the back-projection method and a detailed investigation of the surface rupture in the Xiaoyudong, Sichuan area, which is the junction area of the two main surface ruptures, and of the kinematic fault rupture process. The role of the Xiaoyudong salient is thoroughly investigated, and further implications regarding earthquake rupture processes and regional tectonics are addressed.

Keywords Wenchuan earthquake · Source rupture imaging · Xiaoyudong salient

1.1 Introduction

On May 12, 2008, an earthquake of magnitude 7.9 and with an estimated hypocenter of longitude 103.4° E, latitude 30.1° N, and depth of 14 km occurred in the area along the Longmen Shan fault zone (Fig. 1.1). This event occurred under the tectonic regime of long-term compression and eastward propagation of the plateau. Many geological features indicate that this range is not a typical active convergent mountain belt, such as the fact that it is a young and high mountain, that it has a thickened crust with a very

B.-S. Huang (✉)
Institute of Earth Sciences, Academia Sinica, Taipei, Taiwan
e-mail: hwbs@earth.sinica.edu.tw

C.-P. Chang
Center for Space and Remote Sensing Research, National Central University, Zhongli, Taoyuan, Taiwan

C.-H. Lo et al. (eds.), *Earthquake Geology and Tectonophysics around Eastern Tibet and Taiwan*, Atmosphere, Earth, Ocean & Space,
https://doi.org/10.1007/978-981-15-6210-5_1

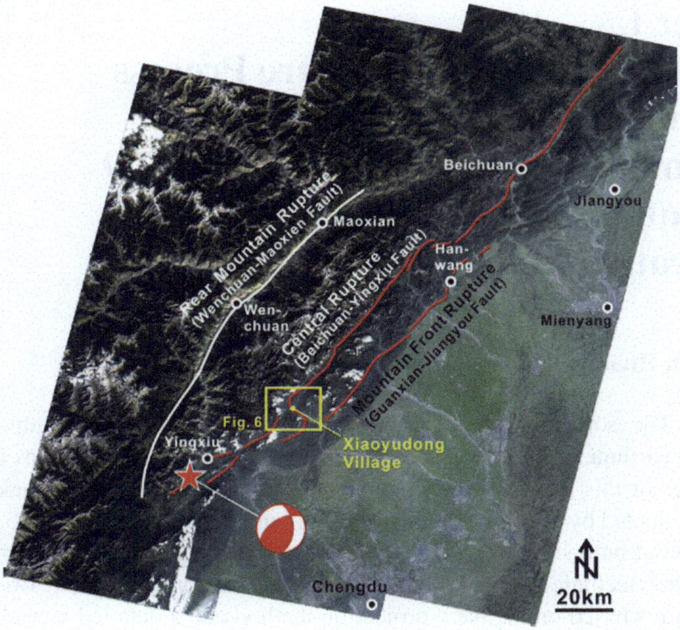

Fig. 1.1 SPOT-5 image acquired after the earthquake. Previous field investigations showed that the Wenchuan earthquake ruptured two imbricate reverse faults along the Longmen Shan fault zone as shown by the red lines. The red star shows the epicenter of the mainshock of the Wenchuan earthquake. The focal mechanism solution is from the United States Geological Survey. The location of Fig. 1.6 is shown by the yellow rectangular frame. Revised from Chang et al. [5]

low GPS shortening rate [50, 51, 53], and that it has no corresponding foreland subsidence [3]. The mapping of its surface faults has revealed that the complicated fault structure appears to comprise several faults and may have simultaneously ruptured parallel faults in the southern region (see for example [7]). The mapped surface displacements range up to approximately 6 m [46].

According to reports by the International Seismological Centre (ISC), there are no historical records of events of magnitude greater than 6.5 specifically within the Longmen Shan area, and no events of comparable size to the 2008 earthquake have been recorded in the general region. The fault plane solution for this earthquake indicates that it is of primarily thrust slip with a right-lateral strike-slip component, corroborating geological observations of associated coseismic surface ruptures. The source mechanism of the Wenchuan earthquake and its aftershocks have been analyzed in several studies [6, 16], all of which reported an almost purely reverse-slip focal mechanism and shallow seismicity. Numerous studies have estimated the slip distribution for the mainshock of the Wenchuan earthquake using teleseismic waveform inversion [8, 9, 20, 31, 33, 40]. All these models showed that the earthquake rupture propagated mainly unilaterally from the hypocenter toward the northeast over a distance of more than 220 km, with an area of large slip near the hypocenter or slightly to its northeast. Based on the field survey and the kinematic source rupture analysis, this earthquake is identified as having ruptured two sub-parallel reverse faults 15–20 km apart, the Beichuan fault (BCF) and the Pengguan fault (PGF) in the Longmen Shan, and also to have exhibited a unilateral ~340 km-long rupture striking NE-SW with thrust and right-lateral components with a high-angle fault dipping to the NW [45]. Previous studies have described the geometry of this earthquake ruptured fault as a curved plane with a decreasing dip from the surface to greater depths [6, 38].

The source rupture processes and coseismic surface rupture of the Wenchuan earthquake are recognized as among the most complicated patterns of recent great earthquakes. In this article, we review some of the Wenchuan earthquake source studies that were carried out by scientists from Taiwan and mainland China with the scientific cooperation of both sides of the Taiwan Strait [5, 15]. In order to clarify how these ruptures developed and their tectonic implications, we describe a detailed investigation of the surface rupture in the Xiaoyudong area, which is the junction area of two main surface ruptures, and the kinematic fault rupture process during this earthquake.

1.2 Data

1.2.1 Seismic Array Observations

Seismic arrays can detect and locate earthquakes and seismic structures with a better signal-to-noise ratio than in the case of a single seismic station [36]. In the early works of source imaging by Huang et al. [15], an approach similar to that described by Ishii et al. [19] was employed, which used data from double regional arrays to image the complex source rupture process of the 2008 Wenchuan earthquake, and thus, the two broadband arrays from Taiwan and northern Vietnam provided similar high-quality regional seismic data to that of the Hi-Net array in Japan [30]. Both are also located at critical azimuths in terms of resolving the rupture processes of the Wenchuan earthquake. They applied a back-projection mapping in which seismograms were stacked for each possible source location to obtain a direct image of the source [10–12, 14, 17]. The stacking procedure sums the energy radiating from the given source points constructively and cancels out all other energy present in the array seismograms. This approach is suitable for not only single array to source imaging, but also for double arrays, which can be implemented under the same stacking procedure. The new combined source image that stacks images approaching from several azimuths is expected to have a higher spatial resolution than that of a single array, which is an advantage in imaging the source rupture.

Seismic data were obtained from two broadband networks (Fig. 1.2): The Broadband Array in Taiwan for Seismology (BATS) and the Vietnam Broadband Seismic Network (VBSN) (Fig. 1.3). BATS is a regional broadband seismic network that

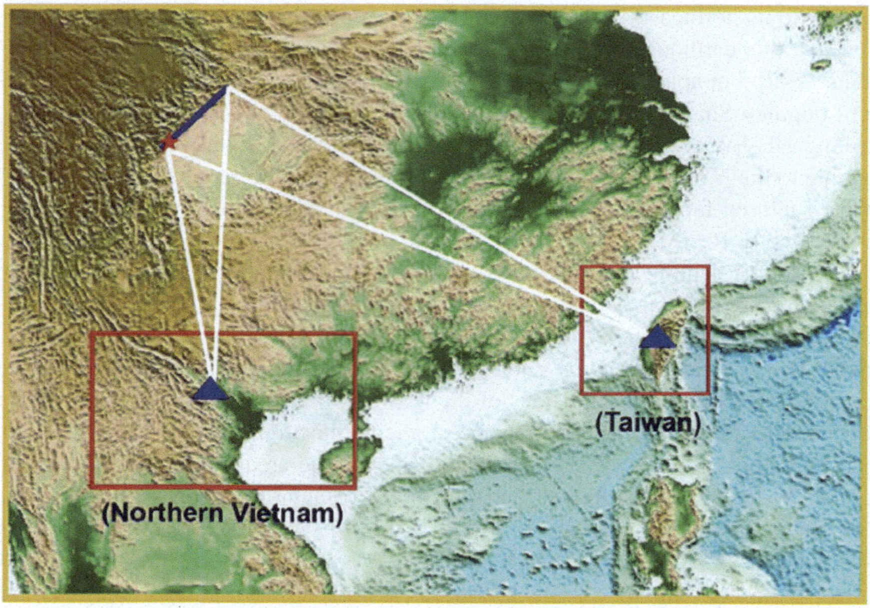

Fig. 1.2 Regional topographical map showing the location of the earthquake fault (thick blue line) of the 2008 Wenchuan earthquake, and the geometry of the two regional seismic arrays used in this study, located in Taiwan and northern Vietnam. The epicenter is indicated by the red star. White lines show the range of paths from the ruptured fault to the array centers (solid blue triangles). Revised from [15]

comprises the Institute of Earth Sciences of Academia Sinica (IESAS) and the Central Weather Bureau Seismic Network (CWBSN) [21]. The network is designed to monitor seismic activity and to determine the regional earthquake source mechanisms in the Taiwan region. For regional events, the integrated BATS seismic network can be considered a dense broadband seismic array, employed to determine the source rupture processes of great earthquakes at regional and teleseismic distances. VBSN is a portable broadband seismic array developed jointly by researchers from Vietnam and Taiwan for imaging and interpreting crust and mantle structures beneath northern Vietnam, including the geodynamic evolution of the Red River Shear Zone (RRSZ), and as an early warning system for earthquakes and tsunamis [13]. This array has a uniform spatial distribution in northern Vietnam and covers the RRSZ and the region of high seismic activity in northwestern Vietnam, with an inter-station spacing of approximately 100 km.

BATS obtained a continuous recording of the 2008 Wenchuan earthquake sequence. The array is located at 16.5°–18.3° from the earthquake and clearly records a direct P-wave. Figure 1.4a shows examples of the P-waves used for source imaging. The P-wave onset of this event is remarkably consistent between the different stations in the array, but later parts of the P-wave train are complicated by multiple overlapping arrivals of seismic energy from different portions of the rupture. For BATS stations

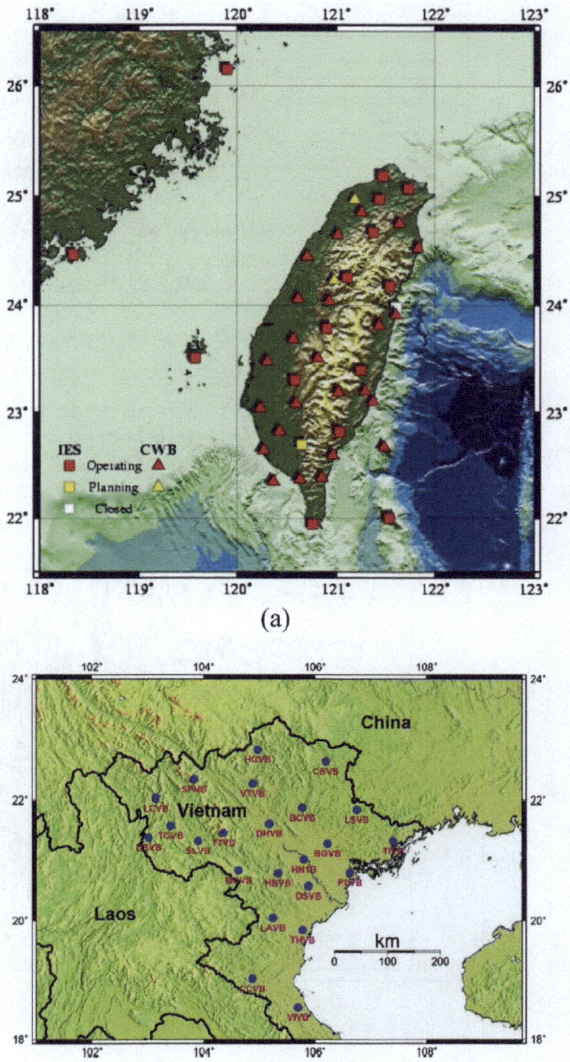

Fig. 1.3 **a** Distribution of seismic stations in Taiwan and on surrounding islands. For more information, see the BATS website. **b** Distribution of seismic stations (blue circles) in northern Vietnam. Station codes are written in red. Revised from [15]

located to the east of the Wenchuan earthquake, the N-S orientation of the ruptured fault plane resulted in a large azimuth gap between the ruptured fault and the BATS array (nearly 10°), as shown in Fig. 1.2; however, because of the regional distance from the earthquake fault to the seismic array, the BATS seismograms still provide sufficient spatial resolution for spatiotemporal imaging of the source processes.

The 2008 Wenchuan earthquake was well recorded by the VBSN array. This array is located at 8.4°–11.5° from the earthquake. The first of these observations were predicted to be Pn-phases by the earth model IASP91 [22]. Figure 1.4b shows examples of the VBSN array seismograms used in the present study. The Pn-wave

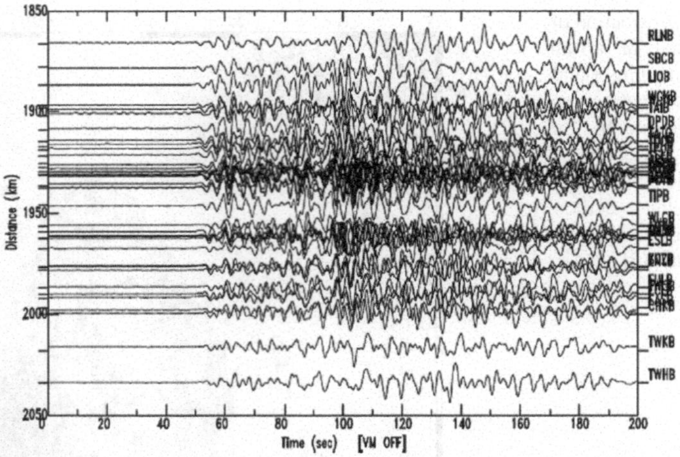

(a)

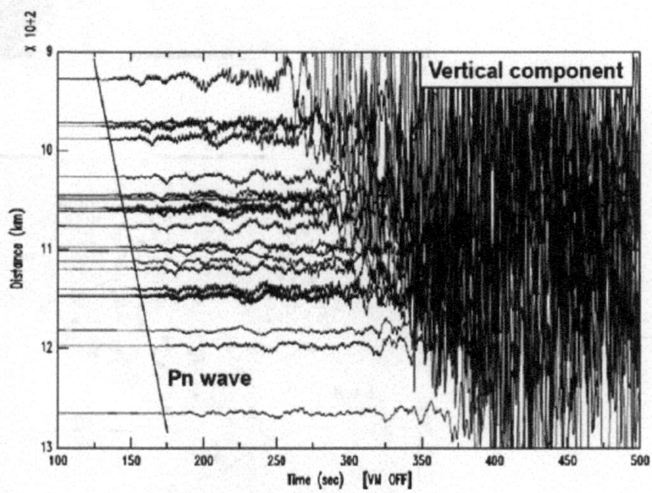

(b)

Fig. 1.4 **a** Aligned vertical-component velocity seismograms of the 2008 Wenchuan earthquake selected from the BATS stations. The seismograms were filtered and normalized individually. The vertical axis shows the distance of each station from the epicenter. After removing high-noise traces and some remote array stations, 45 seismograms were used in this study for seismogram stacking. **b** Distance–travel time diagram for the vertical-component velocity seismograms of the 2008 Wenchuan earthquake selected from the VBSN stations. The seismograms were normalized individually. The horizontal axis indicates the time from the earthquake origin, and the vertical axis shows the distance of each station from the epicenter. Solid lines represent theoretical travel-time curves predicted using Earth model IASP91. Revised from Huang et al. [15]

onset of this event is remarkably consistent between the stations in the array. The azimuth gap within the ruptured fault and this array is almost 14°, which is larger than the azimuth gap of the BATS array (Fig. 1.2). This is because of the shorter distances to the epicenter.

1.2.2 Field Geological Survey of the Xiaoyudong Salient

This earthquake involves a direct manifestation of active crustal shortening and documents its importance in developing and supporting the Longmen Shan Range; however, this crustal shortening did not occur along a single thrust fault. From the distribution of the large aftershocks and reported surface rupture, it appears that the Wenchuan earthquake and its aftershocks caused a slip on two major faults within the imbricate thrust stack (e.g., [44] (Fig. 1.1). The largest fault is the Beichuan–Yingxiu Fault (called the Central rupture or the Central Fault herein). The length of this fault reaches approximately 240 km. Southeast of this fault, a smaller displacement occurred along the Guanxian–Jiangyou Fault (called the Mountain Front rupture or the Mountain Front Fault herein), which has a length of approximately 70 km. The 7 km-long NW-striking Xiaoyudong Fault was also clearly observed between the two main surface ruptures. To clarify how these ruptures developed and their tectonic implications, we reviewed Chang et al.'s [5] detailed investigation of the Xiaoyudong area, which is the junction area of two main surface ruptures.

The nature of the Xiaoyudong Fault is still debated. Many previous studies proposed that this fault is a tear fault linking the Mountain Front and Central ruptures, such that the Longmen Shan fault zone could be considered as a simple imbricate thrust stack [25, 26, 49]; however, field evidence shows that the real situation may be much more complex. Because the nature of the Xiaoyudong Fault plays an important role in understanding how the seismogenic fault developed into two ruptures as well as the tectonic evolution of the Longmen Shan fault zone, in this study, we conduct a detailed investigation that includes fault tracing and kinematic and stress analysis.

Figure 1.5 shows the coseismic fault scarp cutting through Xiaoyudong village. The deformation features in this area include a simple thrust, a hanging-wall collapse scarp, a pressure ridge, a sinister pressure ridge and rupture, a back thrust ridge, and a fault-related fold scarp [5]. Previous studies provided many vertical displacement measurements along the coseismic surface deformation zone of the Wenchuan earthquake because the vertical slip (generally the height of a fault scarp) is the most readily measured element [44]. Chang et al. [5] recorded 3-D measurements of this area to understand the motion behavior of the fault.

Chang et al. [5] recorded the azimuth of the rupture and the azimuths and apparent offsets (including the vertical and horizontal components) of nonparallel linear constructions such as small roads, ditches, or walls that were cut by the rupture. Based on these data, they determined the fault slip vector (uplift, shortening, and strike-slip motion) and the near-surface fault dip-angle using a graphic method. After this detailed survey and analysis, the trace of the Xiaoyudong Fault and its

Fig. 1.5 Xiaoyudong Fault line (dotted line), scarp, and Xiaoyudong village. The right side of the image (in the southern direction) shows the hanging wall. Revised from Chang et al. [5]

kinematic behavior were reasonably well identified. The maximum vertical offset of approximately 1.8 m and the maximum horizontal offset of approximately 2.3 m all appear in the central part of the fault. Based on their analysis, the field results show that the Xiaoyudong Fault is not a continuous fault line, but is composed of a series of fault segments that are not directly connected. It has been suggested that the Xiaoyudong Fault is a tear fault as it is generally a steeply dipping and strike-slip moving structure. However, this may not be correct, and it could be a series of reverse faults (with a left-lateral moving component) generated by NE-SW compression.

In order to understand the tectonic stress and how it influences the regional structure, the fault slickenside striations were measured to conduct a stress analysis of the Xiaoyudong area. These measurement sites are concentrated around the principal surface ruptures of the Xiaoyudong Fault along ruptures located on the hanging wall of the Xiaoyudong Fault and along part of the Central Fault and the Mountain Front Fault. The slickenside striations recorded in the area far from the fault lines or in the rock without clear coseismic deformation were avoided because they may not be directly related to the Wenchuan earthquake. The measured faults were analyzed with a computer using inverse methods [1, 2]. In addition to the slickenside striations on the coseismic genetic fault planes, they also measured the fresh small fault planes in the deformation zone, which resulted from the strain imposed by the Wenchuan coseismic process. By considering the attitudes of the computed principal stress axes relative to the bedding planes at the sites where significant tilting had occurred (especially for the strata dipping more than approximately 30°), it was possible to distinguish the fault events that pre-dated or post-dated the folding. In post-folding faulting, one axis is usually nearly vertical (σ_1, σ_2, or σ_3, depending, respectively, on whether the normal, strike-slip, or reverse faulting mode dominated). In contrast, where pre-folding faulting occurred, this axis is not vertical, but roughly perpendicular to the tilted bedding. The axes of σ_1 of the result are almost horizontal, which means that these slickenside striations were recorded recently (post-folding) and no tilting is needed.

Based on the differences in the types and orientations of the tectonic regimes, two main groups of faulting were distinguished. The first group was reverse in type with a NW-SE trending σ_1. The second group was mainly reverse in type but with a strike-slip component with a NE-SW trending σ_1. A summary of the trends of the σ_1 axes is also shown in the topographic map in Fig. 1.6. The first group (yellow arrows) is distributed along the Central and Mountain Front ruptures, but the second group (red arrows) is concentrated along the Xiaoyudong Fault and its southwestern area. The stresses of the first group, with a compression direction similar to that of the mainshock focal mechanism, can be explained as a result of the SE-moving Tibetan Plateau. The different stresses of the second group might have been caused by the local geometric complexity of the fault system (a salient) that probably determined the local disturbance of the regional stress field during the coseismic rupture. This idea will be discussed in the following section.

1.3 Analysis of the Results

1.3.1 Imaging for Source Rupture Processes

To reconstruct the source rupture on the fault plane, a back-projection approach as described in the previous section was used to image the energy that radiated from the source. This approach uses back-projection mapping, in which seismograms are stacked for each possible source location to obtain a direct image of the source. In practice, because P-wave travel times provide a poor constraint on the source depth, to determine the stacked image, only back-projection to source locations was carried out as a function of latitude and longitude at the hypocenter depth. For back-projection calculations of the Wenchuan earthquake, the model was set as a grid of 200×10 points (400×20 km) in the source area, covering the aftershock region. The travel time from each station to any possible source was computed using a 1D velocity model (IASP91).

For seismic arrays located at regional distances, the different branches of seismic energy that reached the BATS stations from the source area after the major aftershocks of the Wenchuan earthquake were checked, and no clear multiple P-wave arrivals were observed. The predominant branch of seismic energy was identified

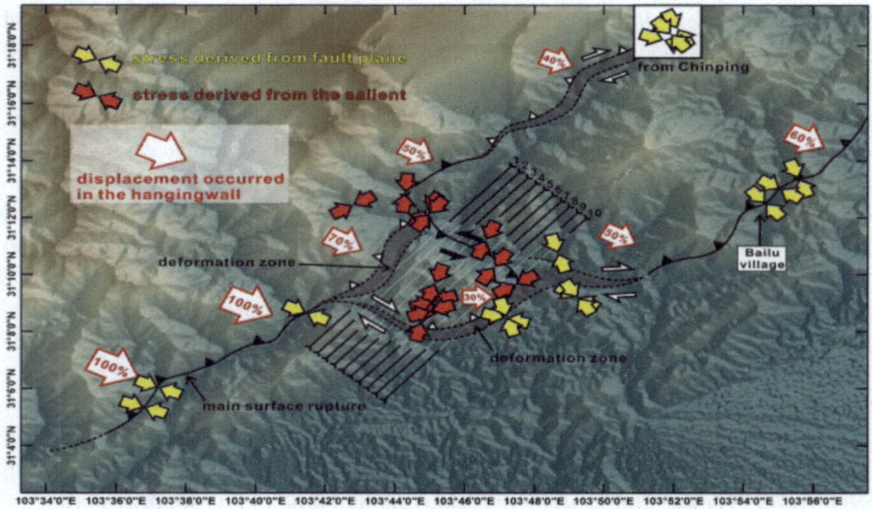

(a)

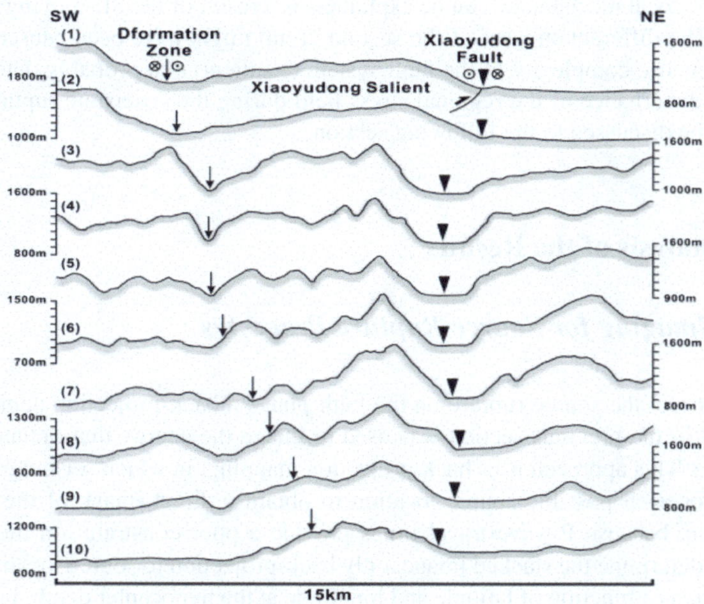

(b)

Fig. 1.6 **a** Summary of stress distribution in and around the Xiaoyudong area. Along the Central and Mountain Front ruptures, the compression trends (yellow arrows) are generally in the NW-SE direction, but those close to the Xiaoyudong area (red arrows) are generally oriented NE-SW. Continuous black lines with small black triangles represent the fault lines with an evident surface rupture. Gray areas bounded by dotted lines with white triangles represent the deformation zones caused by fault movement but without continuous surface rupture. Lines 1–10 on (**a**) are locations of topographic profiles shown in (**b**). The percentages in the red arrows are the estimated fault movements between the Central and Mountain Front faults [24](**b**). Topographic profiles of the Xiaoyudong salient. Revised from Chang et al. [5]

and employed for future array stacking, and all seismograms recorded by the BATS stations (Fig. 1.4a) were back-projected according to the identified P-phase. The initial arrival of the first-time window was assumed to originate from the grid point corresponding to the earthquake hypocenter. Relative time shifts for each time series were calculated based on the theoretical travel times from the station to the grid point using the IASPEI91 model.

We constructed a data volume of snapshots of source grid points at various time steps and obtained images from over the hundred-second duration of the rupture. Based on the single array data, the BATS seismograms generally showed a rupture zone consistent with the epicenter during the early stages of the initiation of the source rupture. The results showed that after a short period of remaining stationary, the rupture zone began to move continuously northeastward away from the epicenter for approximately 80s.

The VBSN array data were analyzed using the same procedure as for the BATS array observations. The source area was also configured using grid points in the same manner as for the BATS data. The first arrivals in all the VBSN array seismograms were Pn-phases (Fig. 1.4b). Relative time shifts for each grid point and time steps were calculated for stacking. The images show the northward propagation of the source rupture, and some artifacts are apparent near the northeastern end of the fault. The discrepancy between the results from the two arrays can be explained by the fact that the source images were individually determined by each array using the seismic energy radiating from a limited aperture, and also because the two arrays have different azimuths to the source.

A joint estimation of the source rupture of the 2008 Wenchuan earthquake using observations from the two arrays may increase the aperture and reduce artifacts, thereby enabling imaging of the rupture processes and stabilized imaging results. In principle, the two arrays could make different contributions to the construction of the source images. For example, the BATS is located perpendicular to the strike of the ruptured fault plane and is sensitive to rupture propagation. In contrast, the VBSN is located at shorter epicenter distances and has a larger azimuth gap compared with BATS.

Figure 1.7 shows snapshots at ten-second intervals of the seismic energy radiating from the fault plane following the initial rupture using both array observations. The results obtained from joint imaging of the two arrays show the same rupture pattern as that obtained from the single arrays, but with significantly less artifacts. Figure 1.7 shows a stable source rupture moving from the epicenter to the northeastern end of the ruptured fault with variable rupture speed. This analysis enabled the reconstruction of the source images and their spatial locations and the determination of the nature of the rupture behavior. This method provides an independent estimate of the rupture duration of the 2008 Wenchuan earthquake, yielding a duration of approximately 100 s, and indicates that the rupture propagated northward along the Longmen Shan Fault over a distance of more than 280 km from the initial rupture point.

To understand the final slip distribution on the fault plane, Huang et al. [15] processed a time-integrated stack that summed the radiated seismic energy for each grid point on the source plane during the rupture process. Imaged snapshots show the

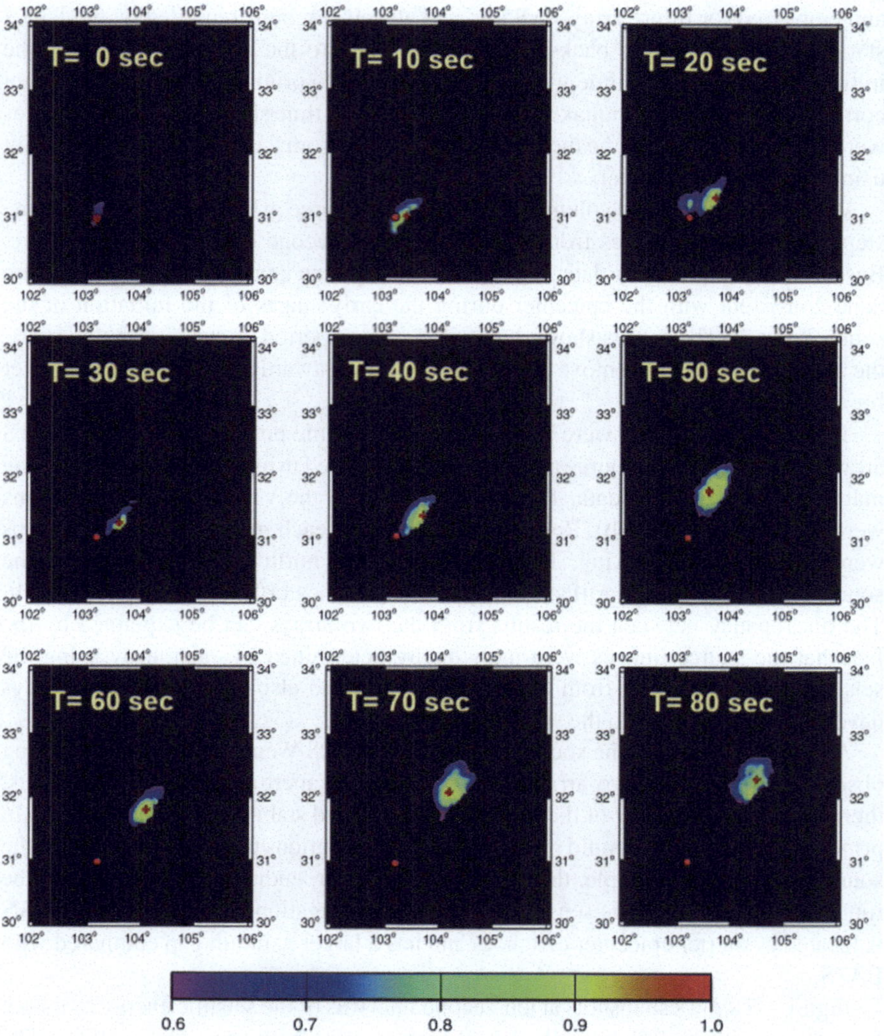

Fig. 1.7 Snapshots of back-projection imaging for rupture processes of the 2008 Wenchuan earthquake using BATS and VBSN observations. Each panel shows the spatial distribution of radiated seismic energy from the fault plane at the stated time after the initiation of the earthquake. The epicenter is indicated by the red solid circle. The red cross represents the peak value in the stacked image. The color area in each panel represents the region in which the radiated seismic energy is greater than 60% of the peak value. The color bar is shown at the bottom. Revised from Huang et al. [15]

stacked seismic energy that radiated from the fault plane, revealing a pattern similar to that of its final slip distribution. The bulk of the seismic energy radiated from a major asperity located nearly 100 km from the epicenter, with a length of 70 km along the rupture fault (Fig. 1.7). A comparison with the results of a previous field investigation [44] reveals that this asperity is located on the Yingxiu–Beichuan Fault near Beichuan County. The snapshots of the source rupture imaged in this present study indicate that the rupturing of this major asperity began 50 s after the initial rupture.

Based on the positions and times of the maximum stack amplitudes for each time window of the arrays (BATS, VBSN, and the joint array), the time–distance relation of the fault rupture slip of the 2008 Wenchuan earthquake was reconstructed to show the propagation of the rupture (Fig. 1.8). The total rupture length was estimated to be approximately 280 km and the duration time as nearly 100 s. Moreover, the results show that the propagation of the fault rupture was mainly unilaterally northeastward with an average rupture speed of approximately 2.8 km/s, although it varied. The rupture began near the southwestern end of the earthquake fault with a rupture velocity of approximately 1.3 km/s; this slow rupture continued for approximately 30 s, at which point the rupture velocity increased to 3.1 km/s for a duration of approximately 50 s. Near the northeastern end of the fault, the fault rupture speed increased to 4.5 km/s before the rupturing finally ceased (Fig. 1.8).

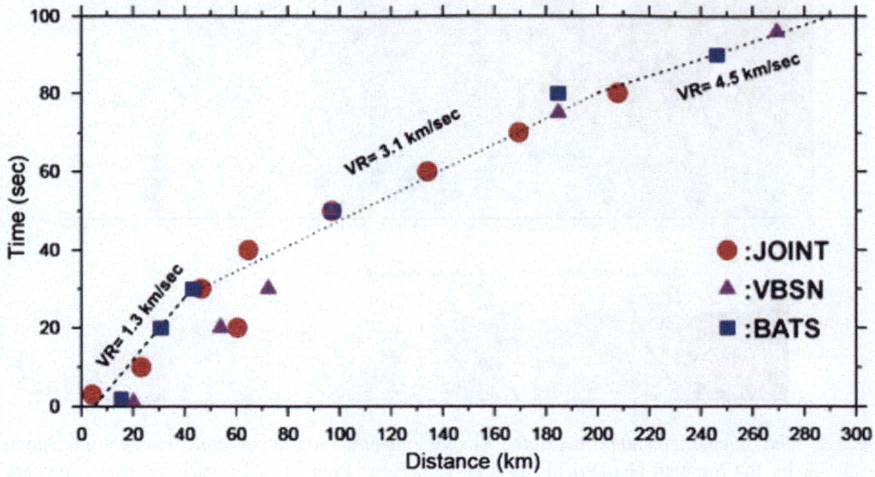

Fig. 1.8 Relation between rupture distance and time during the entire source rupture of the 2008 Wenchuan earthquake. Data were estimated using snapshots from single BATS (blue squares), single VBSN (purple triangles), and JOINT (red solid circles representing the sum of the two arrays). The rupture distance was estimated from the epicenter to the location of the peak amplitude of each snapshot. Dashed lines represent the distance–time relation for the three stages of the rupture process. The average rupture velocity of this earthquake was approximately 2.8 km/s. Revised from Huang et al. [15]

1.4 Proposed Tectonic Model for the Xiaoyudong Area

From a tectonic point of view, the Xiaoyudong salient can be defined as a block that is located south of the Xiaoyudong Fault and occupies a surface area of approximately 100 km^2. Although there is no substantial field evidence, the location and extent of this tectonic salient can still be roughly defined from field investigations. In terms of geomorphology, the Xiaoyudong salient is an independent topographic highland (Fig. 1.6). The strata of this highland are also a little older than those of the adjacent area (Fig. 1.9). These observations allow us to infer that the structures

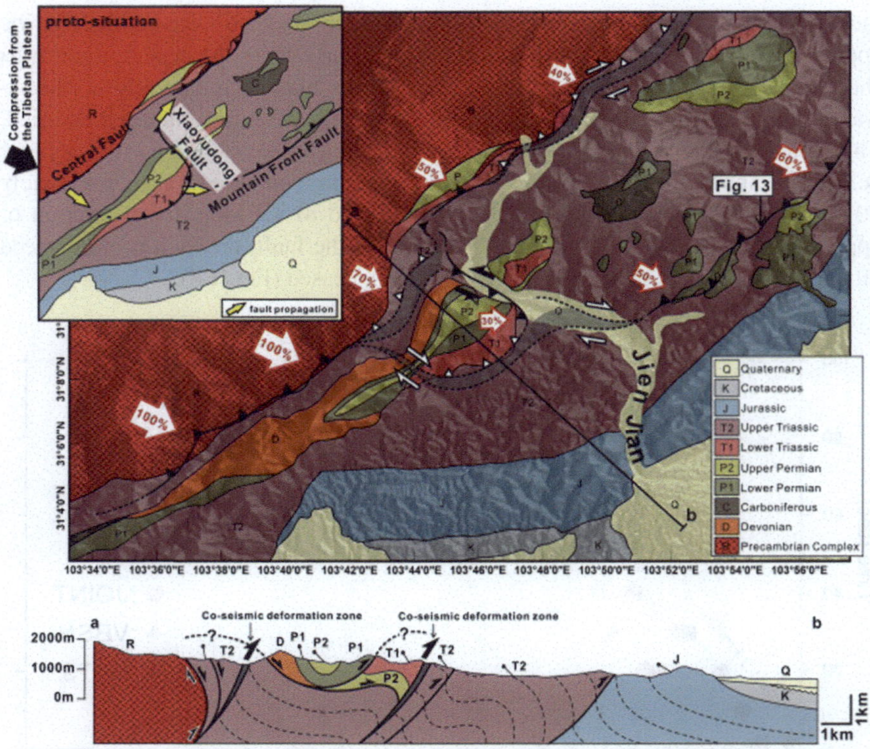

Fig. 1.9 Geological map and profile of the Xiaoyudong area (after an unpublished geological map furnished by the Sichuan Geological Survey). Coseismic fault lines and deformation zones are based on this study. Percentages in the red arrows show the estimated proportion of fault movement between the Central and Mountain Front faults [24]. The upper left rectangular frame shows the simple proto-situation of the regional structure. Yellow arrows indicate the fault propagation direction. In this model, the Xiaoyudong Fault originated as an independent fault east of the Central Fault. This fault did not have large displacement during the Cenozoic, but it has uplifted and folded the Mesozoic and Paleozoic strata by fault propagation. The geological profile shows that the current Xiaoyudong salient is a topographic high associated with a deformed nappe. Revised from Chang et al. [5]

around the Xiaoyudong salient had already been developing for a long time, and that the Xiaoyudong salient is a block that existed before the Wenchuan earthquake.

According to the proposed interpretation, the Xiaoyudong Fault originated as an independent fault east of the Central Fault (Fig. 1.9), although many previous studies have indicated that the Xiaoyudong area did not experience large horizontal displacement during the Cenozoic (e.g., [3]). As a result of compression by the Tibetan Plateau, the materials of this area, including the Mesozoic strata and even the Paleozoic strata on the hanging wall side of the Xiaoyudong Fault, were still uplifted and folded to form a topographic high associated with a syncline, the current Xiaoyudong salient. After this salient formed, the fault propagation behavior and stress delivery of the Xiaoyudong block could change to a certain degree when a large earthquake occurred in the southern area, due to its specific geometry, structure, and material nature.

The Xiaoyudong salient is not a symmetrical topographic high (Fig. 1.6b). The main river divide (the highest range) in the salient is very close to its northern limit. This topographic feature means that the Xiaoyudong Fault is not a passive tear fault but an active fault that has accumulated a large amount of reverse displacement over a long period of time. The S-shape of the stream north of the Xiaoyudong salient (Jien Jian, Fig. 1.9) also suggests that the Xiaoyudong Fault is still active. During the Wenchuan earthquake, the Xiaoyudong Fault had a reverse-oblique left-lateral slip, and this is further evidence that the uplift of the Xiaoyudong salient is still active.

To understand the deeper structures and delineate the three-dimensional geometry of this salient, Chang et al. [5] further reviewed its published geological cross-sections. It has been previously thought that the faulting of the Wenchuan earthquake was superimposed onto a preexisting Mesozoic Orogen. This older deformation provided the initial geometry for the subsequent Cenozoic deformation (e.g., [3]). By comparing the surface ruptures associated with the Wenchuan earthquake with the published geological cross-sections, it was found that almost all of the coseismic surface ruptures belong to two preexisting imbricated reverse faults: the Beichuan–Yingxiu Fault (Central Fault) and the Guanxian–Jiangyou Fault (Mountain Front Fault) [4, 18, 27–29, 43–45]. These two imbricate faults converge at a depth shallower than the hypocenter of the Wenchuan earthquake, which is approximately 20 km deep. For the cross-section domain, Chang et al. [5] proposed that the Xiaoyudong salient can be considered as a wedge-shaped block enveloped by these two imbricate faults and bordered to the north by the Xiaoyudong Fault. As shown in Fig. 1.10, this block is wider at its surface (approximately 10 km) and narrower in its depths, finally converging at a depth of approximately 10 km.

After the Wenchuan earthquake, almost all fault slip distribution models, including those inverted using the earthquake waveform data, GPS data, or Differential Interferometry Synthetic Aperture Radar results, showed an abnormally small slip, approximately 60–70 km north of the Earthquake hypocenter and located near the proposed Xiaoyudong salient (Fig. 1.10, [32, 33, 35, 38]). This discontinuity of the fault slip revealed by independent research is further evidence of the geometric/structural complexity of the seismogenic source (the Xiaoyudong salient) and suggests that it may play an important role in controlling the fault propagation.

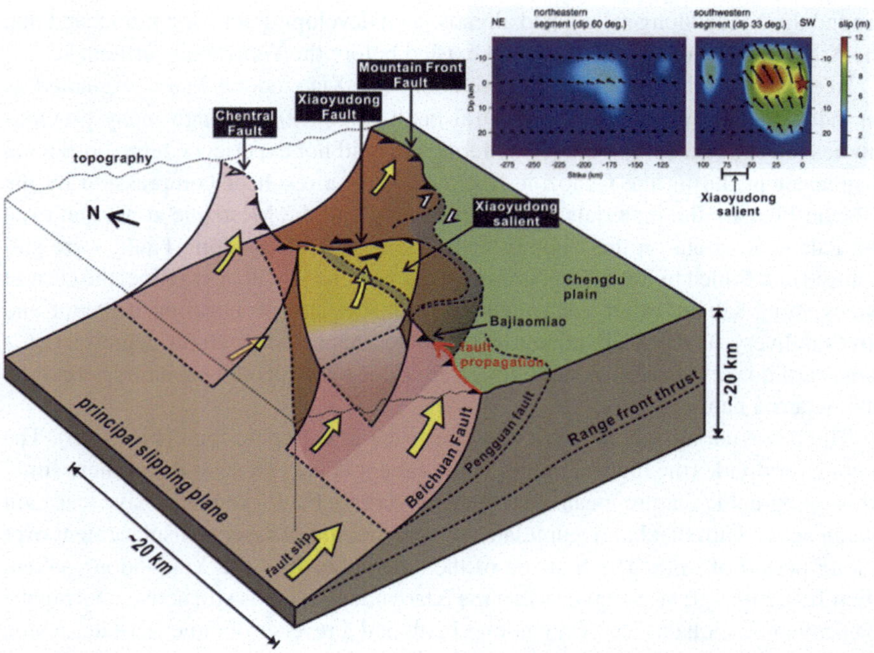

Fig. 1.10 Block diagram showing the possible geometry of the Xiaoyudong salient. Black lines on the surface represent the observed faults, and dotted lines represent the deformation zones (without continuous surface rupture). The underground structures principally follow Burchfiel et al. [3]. Yellow arrows on the fault planes roughly show the amount of relative slip. Red arrows on the fault line indicate the fault propagation direction. The upper right panel is the slip distribution model inferred from the waveform inversion [32]. The small black bar under this model shows the location of the Xiaoyudong Salient. Revised from Chang et al. [5]

The obstructing mechanism of the Xiaoyudong salient is not clear. A possibility is that the Xiaoyudong salient is composed of materials of relatively high strength. When the northward propagating rupture reached the relatively strong Xiaoyudong salient, the rupture itself would have had to divide into two branches. After passing around this salient, the two fault branches continued to propagate northward on separate paths and left the deformation zones around the salient. Along the northern boundary of the Xiaoyudong salient, the fault propagation also dynamically triggered reverse-oblique slip along the NW-SE-striking Xiaoyudong Fault. Evidence for and a detailed discussion of this can be found in Chang et al. [5].

1.5 Discussion and Conclusion

For source imaging, waveform inversions require an indirect procedure to fit observations using a set of theoretical Green's functions computed from an earth model.

The inversion results depend on the accuracy of the theoretical earth model and the computed Green's functions. Various approaches have been proposed to avoid the use of an indirect waveform-inversion procedure, such as direct back-projection of the seismic energy of teleseismic body waves to the source rupture plane to obtain source imaging [47]. In other attempts to improve the spatial resolution of source imaging, the spare Global Seismographic Network data were replaced with dense seismic array observations at teleseismic distances, which provided high-frequency coherent array signals with which to perform high-resolution source imaging [11, 14, 19, 23]. These approaches have been successfully applied to the 2008 Wenchuan earthquake [8, 47].

A similar approach was tested by Huang et al. [15], using new array data from the BATS and VBSN. The main features of the observations of the Wenchuan earthquake are that high-spatial-density broadband seismometers were used at a short distance to the epicenter, and the same as those observations for source rupture are approaching from two arrays at different azimuths. The use of arrays with a high station density enabled us to effectively use the differences in arrival time across each array to determine the approach azimuths of the incident seismic waves at a higher frequency. In addition, observations from a wide range of azimuths (based on two arrays) are useful for reducing artifacts in the source image.

The analysis of Huang et al. [15] resolved a major asperity during the 2008 Wenchuan earthquake (Fig. 1.7). The results indicate that this event had a rupture duration of nearly 100 s, with major asperities radiating energy 50 s after the initiation of the earthquake rupture. The fault rupture began at its epicenter and extended unilaterally to the northeast for about 280 km. Although the average rupture velocity was estimated to be approximately 2.8 km/s, the entire source rupture comprised three segments with different rupture speeds. The rupture began near the southwestern end of the earthquake fault with a rupture velocity of approximately 1.3 km/s. This slow rupture continued for about 30 s, after which time the rupture velocity increased to 3.1 km/s for a duration of approximately 50 s. Near the northeastern end of the fault, the fault rupture speed increased to 4.5 km/s before rupturing ceased. This result is comparable with estimates of the source rupture process made using data from teleseismic and regional distances based on body waveform inversion or array stacking [8, 30, 40, 47, 52].

The Wenchuan earthquake occurred under the tectonic regime of long-term compression and eastward propagation of the eastern Tibetan Plateau; however, many geological features indicate that this range is not a typical active convergent mountain belt. Some of these features are that it is a young, high mountain that has a thickened crust with a very low GPS shortening rate [50, 53], and that it has no corresponding foreland subsidence [3]. Many geologists believe that the crustal thickening that occurred in this area was caused by ductile deformation rather than by thrust faulting or crustal shortening (e.g., [3, 4]). Chang et al. [5] argue that the lower crustal material flowing outward from the center of the plateau is buttressed by the old, strong lithosphere that underlies the Sichuan Basin and pushes up on the overlying crust, maintaining steep topography through dynamic pressure. This hypothesis successfully explains why the upper crust is largely uplifted although there is still very little horizontal shortening at the surface.

Some recent studies based on quantitative structural analysis and that use a balanced cross-section indicate that a large increase in shortening occurs near the range front, and that the structural relief produced by the folds and faults is also closely related to the topography of this front (e.g., [18]). They imply that upper-crustal deformation is the primary mechanism for generating uplift and topography in the foothills of the Longmen Shan Range. This idea obviates the need for lower crustal flow and inflation to produce and maintain the Longmen Shan Range.

There are many different conceptions of the mode of tectonic deformation across the eastern margin of the Tibetan Plateau in the scientific community; however, it is widely agreed that the eastern area has an exceptionally low mechanical strength, inherited from the Mesozoic tectonics of the region (e.g., [48]). The 2008 Wenchuan earthquake involves a direct manifestation of active crustal shortening and is evidence of the importance of active crustal shortening in developing and supporting the Longmen Shan Range. This crustal shortening, however, did not occur along a single thrust fault. From the distribution of the large aftershocks and reported surface rupture, it appears that the Wenchuan earthquake and its aftershocks caused a slip on two major faults within the imbricate thrust stack [44].

The largest fault is the Beichuan–Yingxiu. The length of this fault is approximately 240 km. A smaller displacement occurred southeast of this fault along the Guanxian–Jiangyou Fault, which has a length of approximately 70 km. The 7 km-long NW-striking Xiaoyudong Fault was also clearly observed between the two main surface ruptures. The coseismic surface rupture of the Wenchuan earthquake is one of the most complicated patterns seen in the recent great earthquakes. The Xiaoyudong area, which is the junction of the two main surface ruptures, thus plays an important role in our understanding of how the seismogenic fault developed into two ruptures as well as the tectonic evolution of the Longmen Shan fault zone.

A NW-SE trending cluster of aftershocks can also be clearly observed in the Xiaoyudong area. Many previous studies proposed that this fault is a tear fault linking the Mountain Front and Central ruptures, such that the Longmen Shan fault zone could be considered as a simple imbricate thrust stack (e.g., [25, 26, 49]); however, field evidence revealed by the Wenchuan earthquake shows that the real situation may be much more complex [5, 39]. Some arguments based on fault tracing and kinematic and stress analyses are outlined below.

1. The deformation features in the Xiaoyudong area include a simple thrust, a hanging-wall collapse scarp, a pressure ridge, a sinister pressure ridge and rupture, a back thrust ridge, and a fault-related fold scarp. A tear fault is generally a steeply dipping and strike-slip moving structure; however, detailed survey and analysis results show that the Xiaoyudong Fault is composed of eight southwest dipping thrust fault segments with dip angles ranging from 40° to 60°, and which are not directly connected. The maximum vertical offset of approximately 1.8 m and the maximum horizontal offset of approximately 2.3 m all appear in the

central part of the fault. Because the tear fault absorbs the slip of one thrust or normal fault and transfers it to another fault, the slip amount along a tear fault changes progressively from one to the other. Both the geometry and the degree of slip refute the possibility of the Xiaoyudong Fault being a tear fault. It is in fact a series of reverse faults (with a left-lateral moving component) generated by NE-SW compression.

2. Based on differences in the type and orientation of the stress imposed by the slickenside striations, the faults in the Xiaoyudong area can be distinguished into two main groups. The first group is reverse in type with a NW-SE trending σ_1, distributed along the Central and Mountain Front ruptures. The second group is also reverse in type but with a strike-slip component with a NE-SW trending σ_1, mainly concentrated along the Xiaoyudong Fault and its southwestern area. The stresses of the first group, having a compression direction similar to that of the mainshock focal mechanism, can be explained as a result of the SE-moving Tibetan Plateau. The different stresses of the second group might be caused by a local geometric complexity of the fault system that probably determined a local disturbance of the regional stress field during the coseismic rupture. The fault geometry and kinematics also predict this stress orientation.
3. Although many previous studies indicated that the Xiaoyudong area had no large horizontal displacement during the Cenozoic (e.g., [3]), as a result of compression by the Tibetan Plateau, the materials of this area, including the Mesozoic strata and even the Paleozoic strata on the hanging wall side of the Xiaoyudong Fault, have still been uplifted and folded to form a topographic high associated with a syncline, which is the current Xiaoyudong salient. After this salient formed, the fault propagation behavior and stress delivery of the Xiaoyudong block could change to a certain degree when a large earthquake occurred in the southern area, due to its specific geometry, structure, and material nature. During the Wenchuan earthquake, the Xiaoyudong Fault had a reverse-oblique left-lateral slip, and this proves that the uplift of the Xiaoyudong salient is still active.
4. Almost all the coseismic surface ruptures belong to two preexisting imbricated reverse faults: the Beichuan–Yingxiu Fault (Central Fault) and the Guanxian–Jiangyou Fault (Mountain Front Fault) [4, 18, 27–29, 44, 45]. During the earthquake, when the northward propagating rupture reached the relatively strong Xiaoyudong salient, the rupture itself would have had to divide into the two branches of these two preexisting faults. After passing around this salient, the two fault branches continued to propagate northward along separate paths and left the deformation zones around the salient. The fault propagation also dynamically triggered reverse-oblique slip along the NW-SE-striking Xiaoyudong Fault.
5. "Seismic energy release on a given segment occurs mostly as earthquakes of a size that is characteristic of the segment" [37]. Although the confinement of a fault segment for coseismic rupturing has been widely asserted, ruptures that jump across boundaries into adjacent fault segments are common for large earthquakes. Compared to a single-segment model (rupture confined to a single segment only) estimation of the paleo-earthquake, the multi-segment model contains fewer but

larger magnitude events. The Working Group on California Earthquake Probabilities [42] has contended that many earthquakes in the San Andreas Fault ruptured one or more segments in a single earthquake event. They defined a "cascade model" for simultaneous ruptures on contiguous segments to calculate earthquake occurrence rates in the southern California seismic-hazard analysis. For the extensional Basin-Range province, Wheeler and Krystinik [41] and Pazzopane and Dawson [34] proposed that most of the large historical earthquakes there experienced multi-segment surface rupture. During the Wenchuan earthquake, the Xiaoyudong salient was not an efficient barrier to rupture propagation, but it does play an important role in the paleo-earthquake study and could also influence the paleo-earthquake estimation of the Longmen Shan fault zone.

Acknowledgements This study was funded by the Ministry of Science and Technology, Taiwan, grant MOST 105-2923-M-006-006-MY3 and grant MOST 107-2119-M-001-048 and 108-2119-M-001-019. Thanks to Elsevier for granting permission to use several figures of this study which revised from its original publications.

References

1. Angelier J (1989) From orientation to magnitudes in paleostress determinations using fault slip data. J Struct Geol 11:37–50
2. Angelier J (1990) Inversion of field data in fault tectonics to obtain the regional stress. III. A new rapid direct inversion method by analytical means. Geophys J Int 103:363–376
3. Burchfiel BC, Chen Z, Liu Y, Royden LH (1995) Tectonics of the Longmen Shan and adjacent regions. Int Geol Rev 37:661–738
4. Burchfiel BC, Royden LH, van der Hilst RD, Hager BH, Chen Z, King RW, Li C, Lu J, Yao H, Kirby E (2008) A geological and geophysical context for the Wenchuan earthquake of 12 May 2008, Sichuan, China. GSA Today 18(7):4–11. https://doi.org/10.1130/GSATG18A.1
5. Chang CP, Chen GH, Xu XW, Yuan RM, Kuo YT, Chen WS (2012) Influence of the pre-existing Xiaoyudong salient in surface rupture distribution of the Mw 7.9 Wenchuan earthquake, China. Tectonophysics 530–531:240–250
6. Chen JH, Liu QY, Li SC, Guo B, Li Y, Wang J, Qi SH (2009) Seismotectonic study by relocation of the Wenchuan Ms 8.0 earthquake sequence. Chin J Geophys 52(2):390–397 (In Chinese with an English abstract)
7. Dong S, Zhang Y, Wu Z, Yang N, Ma Y, Shi W, Chen Z, Long C, Meijian AN (2008) Surface rupture and co-seismic displacement produced by the Ms 8.0 Wenchuan earthquake of May 12th, 2008. Sihuan, China: Eastwards growth of the Qinghai-Tibet plateau. Acta Geol Sinica 82:938–948
8. Du HL, Xu LS, Chen YT (2009) Rupture process of the 2008 great Webchuan earthquake from the analysis of the Alaska-array data. China Chin J Geophys 52:372–378 (In Chinese with an English abstract)
9. ERI (2008) Earthquake Research Institute, Univ. of Tokyo, http://www.eri.u-tokyo.ac.jp/topics/china2008/source_eng.html
10. Huang BS (2001) Evidence for azimuthal and temporal variations of the rupture propagation of the 1999 Chi-Chi, Taiwan earthquake from dense seismic array observations. Geophys Res Lett 28:3377–3380

11. Huang BS (2008) Tracking the North Korean nuclear explosion of 2006, using seismic data from Japan and satellite data from Taiwan. Phys Earth Planet Inter 167:34–38. https://doi.org/10.1016/j.pepi.2008.02.004
12. Huang BS, Huang YL, Lee SJ, Chen YG, Jiang JS (2008) Initial rupture processes of the 2006 Pingtung earthquake from near source strong-motion records. TAO 19:547–554
13. Huang BS, Le TS, Liu CC, Toan DV, Huang WG, Wu YM, Chen YG, Chang WY (2009) Portable broadband seismic network in Vietnam for investigating tectonic deformation, the earth's interior, and early-warning systems for earthquakes and Tsunamis. J Asian Earth Sci 36:110–118. https://doi.org/10.1016/j.jseaes.2009.02.012
14. Huang BS, Huang YL, Leu PL, Lee SJ (2011) Estimation of the rupture velocity and fault length of the 2004 Sumatra-Andaman earthquake using a dense broadband seismic array in Taiwan. J Asian Earth Sci 40:762–769. https://doi.org/10.1016/j.jseaes.2010.10.020
15. Huang BS, Chen JH, Liu QY, Chen YG, Xu XW, Wang CY, Lee SJ, Yao ZX (2012) Estimation of rupture processes of the 2008 Wenchuan earthquake from joint analyses of two regional seismic arrays. Tectonophysics 578:87–97. https://doi.org/10.1016/j.tecto.2011.12.026
16. Huang Y, Wu JP, Zhang TZ, Zhang DN (2008) Relocation of the M8.0 Wenchuan earthquake and its aftershock sequence. Sci China Ser D Earth Sci 51(12):1703–1711. https://doi.org/10.1007/s11430-008-0135-z
17. Huang YL, Huang BS, Wang C, Wen K (2004) Numerical modeling for earthquake source imaging; implications for array design in determining the rupture process. TAO 15:133–150
18. Hubbard J, Shaw JH (2009) Uplift of the Longmen Shan and Tibetan plateau, and the 2008 Wenchuan ($M = 7.9$) earthquake. Nature 458:194–197. https://doi.org/10.1038/nature07837
19. Ishii M, Shearer P, Houston H, Vidale JE (2005) Extent, duration and speed of the 2004 Sumatra-Andaman earthquake imaged by the Hi-Net array. Nature 435:933–936
20. Ji C (2008) Preliminary result of the May 12, 2008 Mw 7.97 ShiChuan earthquake. http://www.geol.ucsb.edu/faculty/ji/big_earthquakes/2008/05/12/ShiChuan.html
21. Kao H, Jian PR, Ma KF, Huang BS, Liu CC (1998) Moment-tensor inversion for offshore earthquakes east of Taiwan and their implications to regional collision. Geophys Res Lett 25:3619–3622
22. Kennett BLN, Engdahl ER (1991) Traveltimes for global earthquake location and phase identification. Geophys J Int 105:429–465
23. Krüger F, Ohrnberger M (2005) Tracking the rupture of the $M_w = 9.3$ Sumatra earthquake over 1,150 km at teleseismic distance. Nature 435:937–939. https://doi.org/10.1038/nature03696
24. Kuo YT, Huang MH, Suppe J, Chen YG, Avouac JP, Leprince S, Ayoub F, Kuo YJ (2008) Coseismic ground displacements from sub-pixel correlation for the 2008 Wenchuan Earthquake, Sichuan, China. Eos. Trans Amer Geophys Union 89(53) (Fall Meet. Suppl., Abstract U23B-0057)
25. Li H, Fu XF, Woerd VDJ, Si J, Wang Z, Hou L, Qiu Z, Li N, Wu F, Xu Z, Tapponnier P (2008) Co-seisimic surface rupture and dextral-slip oblique thrusting of the MS 8.0 Wenchuan earthquake. Acta Geol Sinica 82(12):1623–1643
26. Li Y, Huang R, Densmore AL, Zhou R, Yan L, Richardson N, Dong S, Ellis MA, Zhang Y, He Y, Chen H, Qiao B, Ma B (2009) Active tectonics and surface rupture of the Pengxian-Guanxian Fault, Longmen Mountain, Sichuan. China Quat Sci 29(3):403–415 (in Chinese with English abstract)
27. Li H, Si JL, Fu XF, Qiu ZL, Li N, van der Woerd J, Pei JL, Wang ZX, Hou LW, Wu FR (2009) Co-seismic rupture and maximum displacement of the Wenchuan and its tectonic implications. Quart Sci 29(3):387–402 (in Chinese with English abstract)
28. Li H, Si JL, Pei JL, Fu XF, Wang ZX, Li N, Hou LW, Wu FR, Pan JW (2010) Investigating the surface rupture progress of the Wenchuan earthquake (M_s 8.0). Quat Sci 30(4):677–698 (in Chinese with English abstract)
29. Lin AM, Ren ZK, Jia D (2009) C-seismic ground-shortening structures produced by the 2008 M_w 7.9 Wenchuan earthquake, China. Tectonophysics. https://doi.org/10.1016/j.tecto.2009.10.027

30. Mori J, Smyth C (2009) A summary of seismological observations for the May 12, 2008 Wenchuan, China earthquake. In: Investigation report of the 2008 Wenchuan Earthquake, China, grant-in-aid for special purposes of 2008, MEXT, No. 2090002. http://www.shake.iis.u-tokyo.ac.jp/wenchuan/index.html
31. Nagoya (2008) Research Center for Seismology, Volcanology and Disaster Mitigation, Nagoya Univ., http://www.seis.nagoya-u.ac.jp/sanchu/Seismo_Note/2008/NGY8a.html
32. Nakamura T, Tsuboi S, Kaneda Y, Yamanaka Y (2010) Rupture process of the 2008 Wenchuan, China earthquake inferred from teleseismic waveform inversion and forward modeling of broadband seismic waves. Tectonophysics 491(1–4):72–84
33. Nishimura N, Yagi Y (2008) Preliminary results of coseismic-slip for the 2008 Mw 7.9 Wenchuan Earthquake, Southwest China, 2008. http://www.geol.tsukuba.ac.jp/~nishimura/20080512
34. Pazzopane SK, Dawson TE (1996) Fault displacement hazard: a summary of issues and information. U.S. Geol. Surv. Yucca Mountain Report to the U.S. Department of Energy. In: Seismotectonic framework and characterization of faulting and Yucca Mountain, Nevada, Chap. 9, 160 pp
35. Pei SP, Su JR, Zhang HJ, Sun YS, Toksoz MN, Wang Z, Gao X, Liu Zeng J, He JK (2010) Three-dimensional seismic velocity structure across the 2008 Wenchuan Ms 8.0 earthquake, Sichuan. China Tectonophys 491:211–217
36. Rost S, Thomas C (2002) Array seismology, methods and applications. Rev Geophys 40:2-1–2-27
37. Schwartz DP, Coppersmith KJ (1984) Fault behavior and characteristic earthquakes: examples from the Wasatch and San Andreas fault zone. J Geophys Res 89:5681–5698
38. Shen Z-K, Sun J, Zhang PZ, Wan Y, Wang M, Bürgmann R, Zeng Y, Gan W, Liao H, Wang Q (2009) Slip maxima at fault junctions and rupturing of barriers during the 2008 Wenchuan earthquake. Nat Geosci 2:718–724. https://doi.org/10.1038/NGEO1636
39. Tan XB, Lee YH, Chen WY, Cook KL, Xu XW (2014) Exhumation history and faulting activity of the southern segment of the Longmen Shan, eastern Tibet. J Asian Earth Sci 81:91–104
40. Wang W, Zhao L, Li J, Yao Zh (2008) Rupture process of the Ms 8.0 Wenchuan earthquake of Sichuan. China Chin J Geophys 51:1403–1410 (In Chinese with an English abstract)
41. Wheeler RL, Krystinik KB (1992) Persistent and nonpersistent segmentation of the Wasatch Fault Zone, Utah: statistical analysis for evaluation of seismic hazard. In: Gori PL, Hays WW (eds) Assessment of regional earthquake hazards and risk along the wasatch front, Utah. U.S. Geol Surv Profess Pap 1500-A-J, B1-B47
42. Working Group on California Earthquake Probabilities (1995) Seismic hazards in southern California: probable earthquake, 1994–2024. Bull Seis Soc Am 85:379–439
43. Xu L, Rondenay S, van der Hilst RD (2007) Structure of the crust beneath the southeastern Tibetan Plateau from teleseismic receiver functions. Phys Earth Planet Inter 165:176–193
44. Xu XW, Wen XZ, Ye JQ, Ma BQ, Chen J, Zhou RJ, He HL, Tian QJ, He YL, Wang ZC, Sun ZM, Feng XJ, Yu GH, Chen LC, Chen GH, Yu S, Ran YK, Li XG, Li CX, An YF (2008) The M_s 8.0 Wenchuan earthquake surface ruptures and its seismogenic structure. Seismol Geol 30(3):597–629 (in Chinese with English abstract)
45. Xu XW, Yu GH, Chen GH, Ran YK, Li CX, Chen YG, Chang CP (2009) Parameters of coseismic reverse- and oblique-slip surface ruptures of the 2008 Wenchuan earthquake, eastern Tibetan Plateau. Acta Geol Sin 83(4):673–684
46. Xu X, Wen X, Yu G, Chen G, Klinger Y, Hubbard J, Shaw J (2009) Coseismic reverse- and oblique-slip surface faulting generated by the 2008 Mw 7.9 Wenchuan earthquake. China Geol 37:515–518
47. Xu Y, Koper KD, Sufri O, Zhu L, Hutko AR (2009) Rupture imaging of the Mw 7.9 12 (2008) Wenchuan earthquake from back projection of teleseismic*P*waves. Geochem Geophys Geosyst 10:Q04006. https://doi.org/10.1029/2008GC002335
48. Yin A (2010) A special issue on the great 12 May 2008 Wenchuan earthquake (Mw7.9): observations and unanswered questions. Tectonophysics 491:1–9

49. Zhang J, Shen X, Xu Y, Gao Z, Lu X, Pan X (2009) Surface rupture features and segmentation of the M_S 8.0 Wenchuan earthquake. Earthquake 29(1):149–163 (in Chinese with English abstract)
50. Zhang PZ, Shen Z, Wang M, Gan W, Burgmann R, Molnar P, Wang Q, Niu Z, Sun J, Wu J, Sun H, You X (2004) Continuous deformation of the Tibetan Plateau from global positioning system deta. Geology 32:809–812
51. Zhang PZ, Wen XZ, Shen ZK, Chen JH (2010) Oblique, high-angle, listric-reverse faulting and associated development of strain: the Wenchuan Earthquake of May 12, 2008. Sichuan China Ann Rev Earth Planet Sci 38:353–382. https://doi.org/10.1146/annurev-earth-040809-152602
52. Zhang H, Ge Z (2010) Tracking the rupture of the 2008 Wenchuan earthquake by using the relative back-projection method. Bull Seism Soc Am 100:2551–2560. https://doi.org/10.1785/0120090243
53. Zhou R, Li Y, Densmore AL, Ellis MA, He Y, Li Y, Li X (2007) Active tectonics of the Longmen Shan region of the eastern margin of the Tibetan plateau. Acta Geol Sin 81:593–604

Chapter 2
A Panoramic View of the Landslides Triggered by the May 12th, 2008 M_W 7.9 Earthquake in Wenchuan, China

Chong Xu and Xi-Wei Xu

Abstract This paper reviews studies of the landslides triggered by the 2008 M_W 7.9 Wenchuan earthquake as a panoramic view of a vast number of slope failures. It examines (1) descriptions of the disastrous coseismic bedrock landslides, (2) the coseismic landslide database, (3) landslide influence factors and the spatial distribution pattern, (4) spatial landslide susceptibility mapping, and (5) effects of earthquake-triggered landslides on landscape evolution. The Wenchuan earthquake triggered massive large-scale bedrock landslides, which resulted in serious casualties and property damage. At least 197 481 landslides were identified, being distributed within an elliptic area of approximately 110 000 km^2. Their total area and volume were approximately 1160 km^2 and 6–10 km^3. They constitute the largest landslide database related to an individual earthquake ever recorded worldwide. Many factors might have affected these slope failures, such as topography, geology, strong ground motion, and surface deformation related to earthquake magnitude and the properties of the seismogenic faults. Associated with this, areas with the following features are more susceptible to landslides: high slope angles; close proximity to the Yingxiu–Beichuan surface fault rupture; hanging wall of the reverse fault; east-, southeast-, and southward slope aspects; peak ground accelerations larger than 0.2 g; sandstone, siltstone, or granitic underlying rocks; and IX–XI intensity levels. A significant number of methods for earthquake-triggered landslide assessments have been used, including the Newmark method, statistical methods, and combinations of both. The resultant assessment maps provide useful information for the rehabilitation and reconstruction of the affected area. In addition, some researchers have studied the relationships between coseismic landslides and crustal uplift, landscape evolution in the affected area, and the long-term effect of earthquake-triggered landslides on the environment, which will be helpful for promoting the advancement of research on this subject.

C. Xu (✉) · X.-W. Xu
Institute of Crustal Dynamics, China Earthquake Administration, Beijing, China
e-mail: xc11111111@126.com

C. Xu
Key Laboratory of Active Tectonics and Volcano, Institute of Geology, China Earthquake Administration, Beijing, China

C.-H. Lo et al. (eds.), *Earthquake Geology and Tectonophysics around Eastern Tibet and Taiwan*, Atmosphere, Earth, Ocean & Space,
https://doi.org/10.1007/978-981-15-6210-5_2

Keywords Wenchuan earthquake · Landslides · Database · Landscape evolution

2.1 Introduction

It has been more than ten years since the M_W 7.9 earthquake that occurred on May 12th, 2008 in Wenchuan. This event has received much attention from geoscience communities around the world [12, 19, 21, 25, 32, 40, 42]. This huge shock took place in a mountainous area with steep terrain, and the strong ground shaking it caused triggered massive coseismic landslides of an unprecedented number, giving rise to a number of studies [2, 6, 8, 18, 40]. Such large-scale downslope movement of soil and rock involves complex mechanisms and a variety of natural conditions, necessitating further research.

This paper reviews these studies and is broken down into the following sections: various extremely large and disastrous coseismic landslides, the coseismic landslide database, landslide influence factors and the spatial distribution pattern, landslide susceptibility mapping, and the effect of earthquake-triggered landslides on landscape evolution. These contents permit us to create a panoramic view of the landslides triggered by the 2008 Wenchuan earthquake.

2.2 Extremely Large and Disastrous Coseismic Bedrock Landslides

The Wenchuan earthquake-triggered landslides directly caused approximately 20 000 deaths, which account for approximately 25% of the total death toll from the event [43, 45]. A portion of these was caused by the huge and disastrous bedrock landslides that took place during the shock. Table 2.1 lists eleven typical examples of extremely large landslides. For instance, the Wangjiayan landslide that occurred at Beichuan buried approximately half of the old Beichuan County and resulted in the deaths of more than 1600 residents (Fig. 2.1). The elevations of the back edge and front edge of this landslide are approximately 940 m and 650 m, respectively. Its runout distance is approximately 550 m with a sliding direction roughly N80° E. It is located at the hanging wall of the seismogenic fault with a source area at approximately 400 m from the fault. During the earthquake, the area where this slope failure occurred suffered shaking of XI intensity [28]. It was reported that this large landslide has been progressively active since 1985 due to rainfall and human activity. There were many ground fissures on the slope before the earthquake, and the quake finally triggered this large-scale landslide [44].

The largest landslide triggered by the Wenchuan earthquake occurred at Daguangbao, located in Gaochuan, Anxian County (Fig. 2.2). It resulted in the collapse of a 3040 m high mountain and the accumulation of material in a valley at a 1380 m elevation. The total length of the landslide was approximately 4650 m.

Table 2.1 Eleven extremely large landslides triggered by the 2008 Wenchuan quake

Name	Location	Area (10^4 m^2)	Volume (10^6 m^3)	Lithology
Wangjiayan	31° 49′ 34″ N, 104° 26′ 53″ E	12.5	4	Siltstone, sandstone, and siliceous rock
Daguangbao	31° 38′ 28″ N, 104° 06′ 41″ E	720	1000	Limestone, dolomite
Tangjiashan	31° 50′ 22″ N, 104° 25′ 58″ E	72	20	Siltstone, sandstone, and siliceous rock
Niujuangou	31° 02′ 43″ N, 103° 26′ 58″ E	49	4	Fine-grained diorite
Wenjiagou	31° 33′ 11″ N, 104° 09′ 05″ E	300	100	Limestone, fine sandstone, mudstone
Yuejiashan	31° 26′ 52″ N, 103° 58′ 45″ E	90	20	Granodiorite
Donghekou	32° 24′ 30″ N, 105° 24′ 30″ E	104	12	Carbonaceous and siliceous slate, phyllite, and dolomitic limestone
Woqian	32° 18′ 23″ N, 104° 57′ 48″ E	67	15	Limestone, dolomite, marble, phyllite
Xiejiadianzi	31° 18′ 03″ N, 103° 50′ 19″ E	29	0.45	Plagioclase and amphibole schist
Taihongcun	31° 58′ 30″ N, 104° 35′ 49″ E	60	4.8	Shale, siliceous rock, dolomite, and limestone
Beichuanxinzhong	31° 49′ 45″ N, 104° 27′ 34″ E	12	2.5	Thick layer limestone

The landslide covered approximately 7.20×10^6 m^2 with a volume of approximately 1×10^9 m^3. It occurred in an X intensity area on the hanging wall of the seismogenic fault approximately 4.5 km away. As a typical landslide, it was composed of a giant potentially unstable block caused by the intersection of a rock layer and two groups of structural planes, sliding in the N80° E direction in the downslope motion. Several studies [1, 11, 28, 46] have shown that the accumulated material of this landslide largely maintained the original structure of the bedrocks with a moderate disintegration of its body, forming a deposit with a length of 4.2 km and a width of 2.2 km. Several other large-scale slope failures occurred downstream of the Daguangbao landslide, e.g., the Laoyingyan landslide (Fig. 2.2) with a volume of approximately 15×10^6 m^3, approximately 1 km from Daguangbao.

Fig. 2.1 The Wangjiayan landslide triggered by the 2008 Wenchuan earthquake

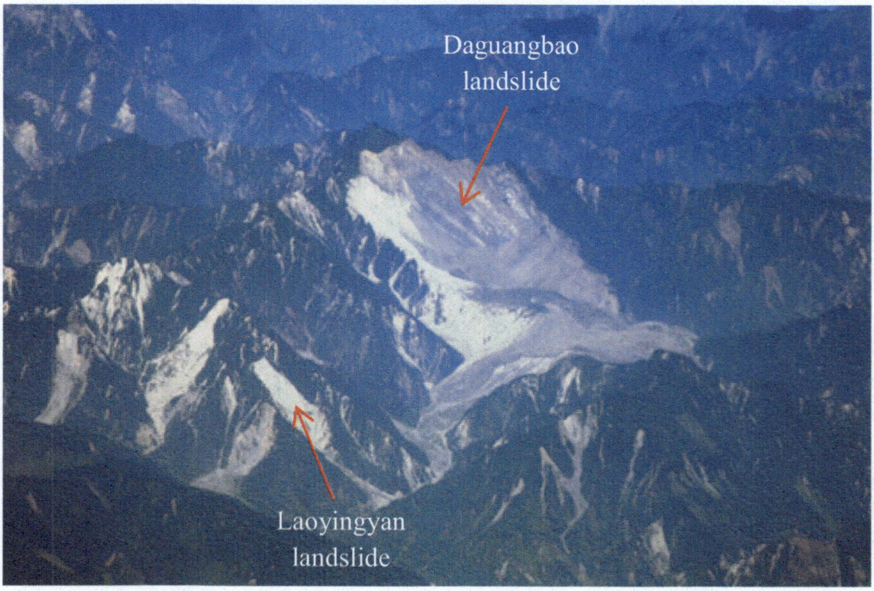

Fig. 2.2 The Daguangbao and Laoyingyan landslides. Photo taken by Huang and Fan [10]

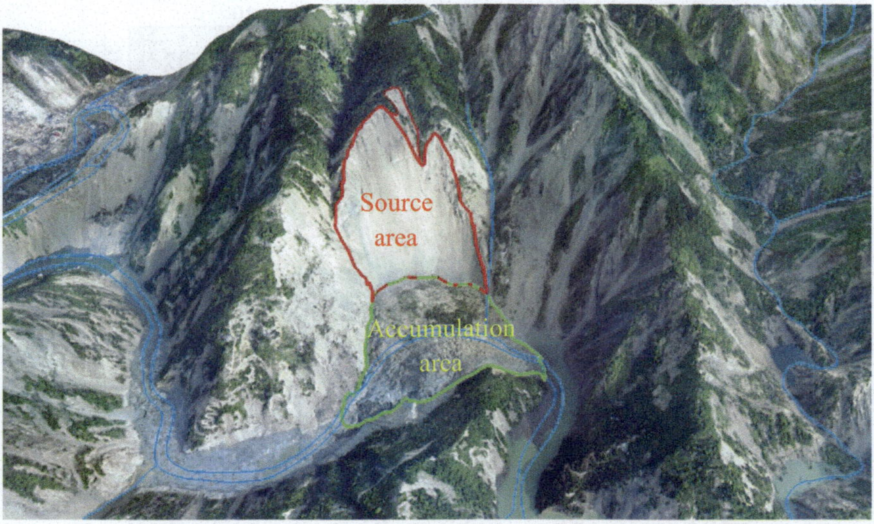

Fig. 2.3 The Tangjiashan landslide

Another large landslide, Tangjiashan, occurred on the right bank of the Jianjiang River. It completely blocked this river, resulting in the largest and most dangerous dammed lake formed by the earthquake (Fig. 2.3). The back and front edges of this landslide are approximately 1260 m and 640 m high, respectively. Its sliding distance was approximately 1150 m in a north-northwest direction. The landslide dam along the river is approximately 800 m long with a maximum width of 610 m and a maximum height of 120 m. This landslide is also located at the hanging wall of the seismogenic fault, approximately 2 km from the fault itself, within an area that experienced XI degree intensity damage. Numerical simulations have shown that the maximum speed the landslide reached was 22 m/s and that there were approximately four stages during the failure: (1) the earthquake force caused progressive destruction of the slope body, (2) the slide body moved downslope at a high speed, (3) this impacted on the slope of the opposite bank causing disintegration, and (4) this blocked the Jianjiang River thereby damming the lake [39].

2.3 Database of Landslides Triggered by the Wenchuan Earthquake

2.3.1 Landslide Inventory

Immediately after the 2008 Wenchuan earthquake, many satellite companies and Chinese government departments quickly organized image shooting of the affected

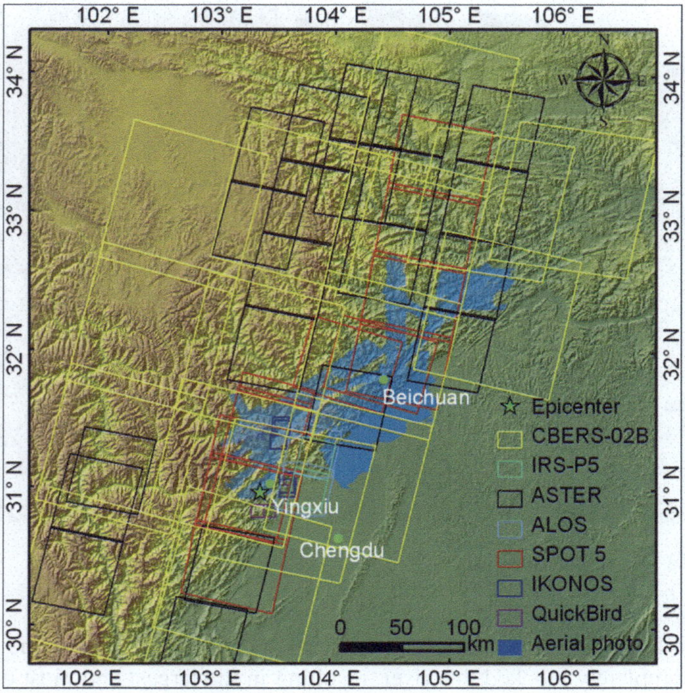

Fig. 2.4 Coverage of post-earthquake images related to the Wenchuan quake

area, and many available satellite images and aerial photographs were released. Based on the visual interpretation of these pictures, Xu et al. [40] constructed a detailed database of the landslides triggered by the earthquake and verified part of it by field investigations. Figure 2.4 shows the coverage of the satellite images and aerial photographs used in their work. Post-earthquake satellite images were from SPOT 5, CBERS02B, IKONOS, ASTER, IRS-P5, QuickBird, and ALOS with 1, 2, 2.4, and 5 m resolutions. In addition, in the remote areas from the epicenter where high-resolution images were not available, satellite images from Google Earth were used to delineate hundreds of landslides (Fig. 2.4). Pre-earthquake images were obtained from SPOT 5, Landsat ETM+, and Google Earth [40]. Figure 2.5 shows pre- and post-earthquake QuickBird images along the Minjiang River in the affected area. A comparison of the images shows that the Wenchuan earthquake triggered massive landslides that destroyed significant amounts of forest and vegetation.

The inventory mapping shows the Wenchuan earthquake triggered at least 197,481 landslides, distributed in an area of approximately 100 000 km^2 and with a total occupation area of approximately 1 160 km^2 (Fig. 2.6). Most of the landslides (196 007 landslides, with a total area of 1150.622 km^2 and accounting for 99.25% and 99.19% of the respective totals) occurred in a smaller area of approximately 44 031 km^2. The number of landslides triggered by the Wenchuan event is the largest amount triggered by an individual earthquake or an earthquake sequence ever reported. Based

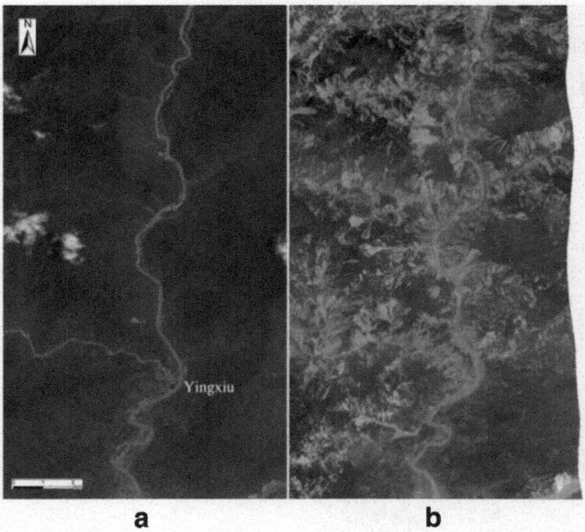

Fig. 2.5 QuickBird images along the Minjiang River, which was struck by the Wenchuan earthquake. **a** Pre-earthquake image mainly composed of forests in green. **b** Post-earthquake image mainly composed of coseismic landslides in light gray

on the "landslide volume–area" power law relationship fitted from the volume of each deposit of the 1415 landslides triggered by the earthquake based on pre- and post-quake DEMs, the total volume was calculated to be approximately 10 km^3 in the source areas [38].

After the Wenchuan earthquake, many research teams released inventories of landslides triggered by this event. Table 2.2 lists twelve typical inventories from the literature. These inventories provide important knowledge for understanding the spatial patterns, hazard analysis, and mechanisms of the earthquake-triggered landslides; however, most of them have some shortcomings, such as incompleteness, positive and negative errors from automatic extraction, only using points to locate landslides rather than polygons to delineate landslide boundaries, and delineating many individual landslides into one landslide. These shortcomings have resulted in significant differences between them, which are likely related to the research aims and characteristics of the researchers. This section presents the inventory of the Wenchuan earthquake-triggered landslides prepared by Xu et al. [40], since it strictly follows the criteria for landslide inventories [30] and is the most detailed and complete of the inventories, even including those of small-scale and those that occurred in areas that are distant from areas of human activity.

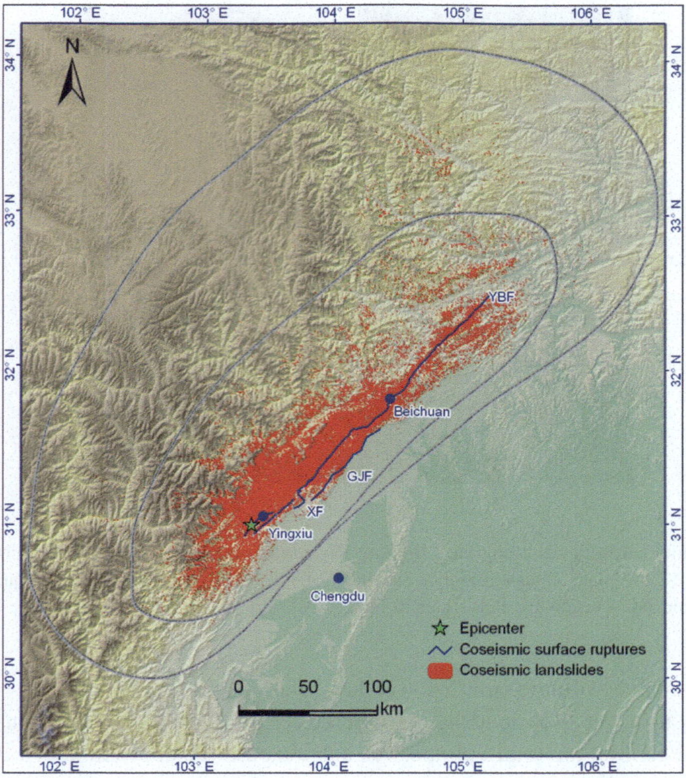

Fig. 2.6 Distribution of landslides triggered by the 2008 Wenchuan earthquake. The larger ellipse is the approximate limited area and the smaller one represents the primary distribution area. YBF: Yingxiu–Beichuan fault; GJF: Guanxian–Jiangyou fault; XF: Xiaoyudong fault

2.3.2 Landslide Characteristics and Spatial Distribution

The Wenchuan earthquake-triggered landslides are characterized by their extremely high number, high-density distribution, large-scale individual slope failure, and broad distribution and clustering. Large-scale bedrock landslides tend to have long runout distances, high speeds, short duration times, large energy, rock fragmentation, complex mechanisms, and the presence of a "throwing effect", and thus often cause serious disasters. The spatial distribution of these landslides exhibits a nearly elliptical shape, with a long axis roughly the same as the strike of the seismogenic fault.

The majority of the landslides triggered by the Wenchuan earthquake were mostly concentrated around the surface rupture zones, implying a strong control by the causative structure [34]. Most of the landslides occurred to the northeast of the epicenter, extending some distance in that direction, consistent with the unidirectional rupturing mechanism of the seismogenic fault. Alternatively, they are mainly

Table 2.2 Database of landslides triggered by the Wenchuan earthquake

No.	Method	Type	Number	Landslide area (km^2)	Source
1	VI	Plg	197481	1160	[40]
2	VI	Pt	11308		[8]
3	VI	Pt	16704		[9]
4	VI	Plg	43842	632	[14]
5	VI	Pt, Plg	13085	418.85	[20]
6	VI	Pt	60104		[6]
7	AE	Plg	73367	565.8	[18]
8	VI	Plg	48007	712	[31]
9	VI	Plg	56000	811	[2]
10	VI	Plg	5154	1139	[7]
11	AE	Plg	57150	396	[16]
12	VI	Plg	6727	54.6	[23]

Note pt point; *plg* polygon; *VI* visual interpretation; *AE* automatic extraction

distributed on the hanging wall of the seismogenic fault, especially in the section between the town of Yingxiu and Beichuan County.

The density of the landslides shows obvious differences along the fault. The landslides along the section from Yingxiu to Beichuan County, which is dominated by thrust, are much more numerous than those along the section north of Beichuan dominated by strike-slip. There are several high-density landslide areas, such as the hanging wall area of the Yingxiu–Beichuan fault between Yingxiu and Beichuan County, the area between the Yingxiu–Beichuan surface rupture and the Guanxian–Jiangyou surface rupture, and the area northeast of Beichuan crossed by the surface rupture. Furthermore, in the hanging wall area far from the seismogenic fault, there are also dense coseismic landslides, mainly along the rivers. The long strip-like area between the Yingxiu–Beichuan surface rupture and the Guanxian–Jiangyou surface rupture also has an area of high landslide density, though it is less dense than that on the hanging wall of the Yingxiu–Beichuan fault [6, 8, 40].

2.4 Landslide Susceptibility Mapping and Hazard Assessment

Earthquake-triggered landslide assessments can provide information about which locations are more susceptible to landslides and which are more stable, which is helpful for the reconstruction and mitigation of subsequent hazards in the affected area. Several important factors play a role in this, such as sample selection, models, and the distribution of the landslides. After the Wenchuan earthquake, quite a few methods were used, primarily the Newmark method and appropriate statistical methods. The Newmark method has two advantages: it does not need real landslide distribution data, nor does it need to consider the physical mechanism. For example, Wang et al. [26] proposed a simplified Newmark method for an emergency hazard assessment of landslides triggered by the Wenchuan earthquake in the eleven counties most severely struck by the quake. They used the instantaneous peak ground acceleration, engineering geological lithology grouping, and topographic data to develop a model for landslide hazard assessment based on the ArcGIS software. Godt et al. [5] employed topographic and geologic data and strong ground shaking to prepare a map of the landslide susceptibility of the area struck by the Wenchuan earthquake when no landslide inventory was available at that time. The resulting accuracy of the Newmark method, however, is often unsatisfactory because a significant number of the regional parameters it needs are difficult to obtain, and thus, the method often relies on generalized values [29].

Statistical methods are more suitable than the Newmark method for researching random processes in nature, including landslides, and have thus been widely used in the susceptibility mapping and hazard assessment of the Wenchuan seismic region. These methods need landslide inventories, which are difficult to obtain directly after earthquakes, and thus the completion of the mapping by this approach may take additional time. Nevertheless, the resultant map can be quite consistent with the distribution of real landslides. For example, Xu et al. [33] used a weight of evidence model and a simple binary statistical method to map the earthquake-triggered landslide hazard in the Qingshui river area stuck by the Wenchuan quake, where at least 2,321 landslides have been identified. A comparison shows that the success rate of the mapping is 71.8%, which is a moderate degree of accuracy. Song et al. [22] proposed a hybrid method based on the Bayesian network and obtained an accuracy of approximately 93% in the assessment of a small study area of Beichuan County.

Selecting training samples from a landslide inventory is important for the quality of landslide susceptibility mapping. To address this issue, Xu et al. [35] designed three strategies for landslide sampling including centroids of large landslides, centroids of all landslides, and randomly selected points from all landslide polygons. In addition, support vector machine (SVM) models that consider four kernel functions were used to check the importance of kernel functions in SVM modeling. The results showed that radial basis function based SVM modeling with randomly selected samples from a landslide polygon inventory is most successful. In another study by Xu et al. [36], four landslide susceptibility models (bivariate statistics, logistic regression, artificial neural networks, and SVM) were used to study a small watershed strongly struck by the Wenchuan earthquake. The results showed that the logistic regression generates the highest success rate (80.34%), which is a counterexample to the contention that computing technique-based models usually have better performance than traditional models. Based on the back-propagation artificial neural network model, Li et al. [13] carried out landslide susceptibility assessments for earthquake- and rainfall-triggered landslides in an area of Qingchuan County struck by the Wenchuan earthquake. Their results showed that the mapping of rainfall-triggered landslide susceptibility by considering slope angle, elevation, slope height, and distance to rivers has the best success rate, and the assessment of earthquake-triggered landslide susceptibility by considering slope angle, elevation, slope height, distance to the rivers, and distance to faults has the best performance. Their study provides a reference for selecting influence factors in the assessment of rainfall- and earthquake-triggered landslide susceptibility. Wang et al. [27] combined the Newmark displacement model and logistic regression to develop a mechanism-based landslide occurrence probability model and adopted it for the use in the Wenchuan and Beichuan Counties. The success rates of the model in these two areas were 79.7% and 75.3%, respectively, which are acceptable.

As the Wenchuan earthquake affected a significant area where numerous landslides were triggered, the landslide susceptibility and hazard assessment of the entire affected area based on complex statistical methods is very time-consuming and difficult to implement. Only a few studies of this subject based on simple statistical methods have been carried out. For example, Xu et al. [41] used the weight index model, which is a bivariate statistical approach, to perform a landslide susceptibility assessment with a large inventory of landslides triggered by the earthquake. In addition, eight cases of different combinations of landslide impact factors were considered in their work to reveal the probable effect of impact factors on the landslide susceptibility assessment and the most suitable combinations of influence factors for landslide susceptibility mapping. It was found that the combination of fault rupture, topographical, and geological factors was the best, which yielded a considerably high success rate of up to 90% (Fig. 2.7).

2.5 Geomorphologic Evolution of the Affected Area

Earthquake-related crustal uplift and broad coseismic landslides play a couple of contradictory roles in landscapes. Crustal uplift builds topography, whereas coseismic landslides destroy it. The confrontation between these two factors has received much attention in recent years. The Wenchuan earthquake caused significant crustal uplift and thus promoted the orogenic effect in the area [42]. The wide-distribution, high-density, and large-scale coseismic landslides triggered by the earthquake reduced the influence of crustal uplift on the regional topography. Therefore, this earthquake provides an unprecedented chance to examine this subject.

Based on the information obtained from the automatic extraction of post-earthquake satellite images and subsequent manual checking, Parker et al. [18] established a landslide inventory that contains 73,367 landslides with a total area of 565.8 km^2. A landslide "area–volume" power law relationship was used to calculate the total volume of the landslides as 5–15 km^3. The crustal uplift caused by the earthquake was derived from the coseismic deformation based on a combination of C- and L-band Synthetic Aperture Radar offset data inverted from ascending and descending tracks [3] which is only 2.6 ± 1.2 km^3. This unexpected result, i.e., the volume of coseismic landslides being larger than the crustal uplift, shows that the Wenchuan earthquake did not play an orogenic role, but may have led to a reduction in the topography [18]. This is interpreted as the crustal uplifts having been caused

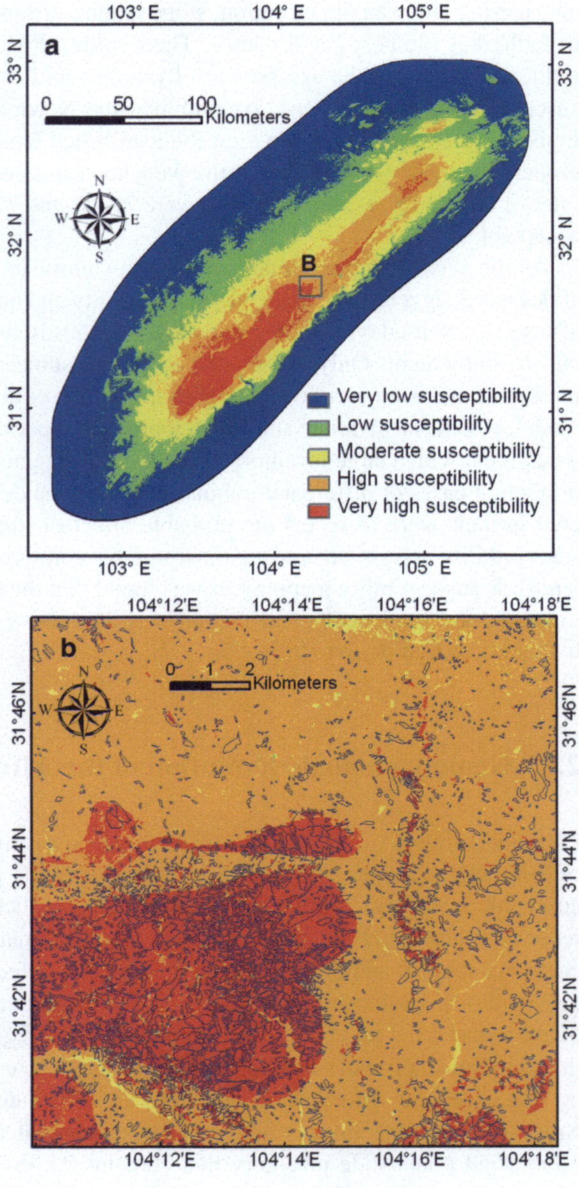

Fig. 2.7 **a** Mapping of Wenchuan earthquake-triggered landslide susceptibility. **b** Enlargement of (**a**)

by more frequent small earthquakes that triggered only a small number of landslides, thus playing a positive role in orogeny. Another possibility is that the relationship between crustal uplift and coseismic landslides caused by earthquakes of similar magnitudes is irregular and cannot be calculated using multiple earthquake cycles. In addition, the seismogenic fault of the Wenchuan earthquake is highly partitioned, and the relationship between its crustal uplift and landslide erosion is different in the thrust fault and strike-slip fault segments. The Parker et al. [18] study has received extensive attention and discussion. For example, based on a visual interpretation of the high-resolution remote sensing images, Xu et al. [37] prepared a more objective and complete landslide inventory for the Wenchuan earthquake and asserted that Parker et al. [18] delineated multiple small landslides as one large landslide in many cases, which undoubtedly would have greatly increased the total volumetric results of the coseismic landslides. Thus, the total volume of the coseismic landslides may have been overestimated.

Using a method similar to Parker et al. [18], Li et al. [16] updated the findings on the coseismic landslide erosion caused by the Wenchuan earthquake. Their results showed that the total volume of landslides triggered by the Wenchuan earthquake was approximately 2.8 km^3, which is much smaller than that found by Parker et al. [18]. This new result is roughly equivalent to the crustal uplift (2.6 ± 1.2 km^3). Assuming the landslide materials can be moved out of the area by rivers during earthquakes, the crustal uplift caused by the quake could be substantially offset by the coseismic landslides.

Li et al. [16] also predicted the crustal uplift of a reverse fault earthquake by considering the earthquake magnitude and the parameters of the rupture surface. Combined with the power law relationship between the total volume of coseismic landslides and earthquake magnitude, the equilibrium relationship between the crustal uplift and the coseismic landslides was obtained [16]. The results are in good agreement with the 1999 Chi-Chi earthquake of Taiwan, China. They also suggested that moderate earthquakes (M_w 6–7) might play a more important role in orogenesis than larger earthquakes because the decrease in coseismic landslides is more significant than the decrease in the amount of crustal uplift for earthquakes of smaller magnitudes [16]. This result is essentially consistent with the analysis and inference of Parker et al. [18]. Nevertheless, many other factors, such as lithology, climatic conditions, focal depth, and tectonic settings, have not been taken into account. Therefore, there is considerable uncertainty and further worthwhile research on this subject.

In the study of landscape changes caused by the Wenchuan earthquake, the time required for landslide deposits to be moved out of the area is an important factor that needs to be considered. Wang et al. [24] assessed the time taken for fine particles (size <0.25 mm) from coseismic landslides to be moved out of the earthquake-affected area based on the daily suspended sediment discharge from 16 major rivers between 2006 and 2012. The results showed that the amount of suspended sediment discharged after the earthquake is approximately 3–7 times that of before the earthquake. Nevertheless, the output of the rivers in the 68,719 km^2 area from 2008–2012 was only approximately 92.5 million tons, which is much less than the estimated amount of fine-grained materials (about 480 million tons) from the Wenchuan earthquake-triggered landslide as determined by deposit grain-size distributions. Their study also used post-earthquake sediment export rates to estimate the maximum time for fine particles to be moved out over 100 years. The study demonstrates that the density of landslides and runoff intensity have an important influence on the length of time the river systems are affected by coseismic landslides.

Li et al. [15] evaluated the relationship between the locations of the coseismic landslides and rivers in the Wenchuan quake area. Based on the landslide number, area, and volume, the connectivity between the coseismic landslides and rivers was quantified. The results showed that approximately 43% of the total landslide volume was connected with the river system in the affected area. The output of fine particles (size <0.25 mm) from the Wenchuan earthquake landslides was positively correlated with the landslide density on a basin scale; however, whether connectivity is considered or not, the statistical correlation is indistinguishable. The weaker than expected connectivity may have an effect on suspended sediment production due to the high mobility of fine particles on slopes, even though they may have had no connection with the river system. The transport of large particles, which accounts for more than 90% of the total landslide volume, may be strongly affected by the location of landslides. This study is helpful in understanding the mechanism of landslide material migration in the affected area.

Earthquake-triggered landslides will undoubtedly increase the regional topography erosion rate in the short term (one year to several decades) [4]. To evaluate whether earthquake-triggered landslides can contribute significantly to landscape erosion over a longer period of time (over thousands to millions of years), Li et al. [17] simulated the volumes of landslides triggered by earthquakes with different magnitudes and calibrated the results using the coseismic landslide data from the Wenchuan and Lushan earthquakes. The results showed that the long-term coseismic landslide erosion rate has an equivalent magnitude to the long-term regional erosion rate, which is 0.5–1 mm/year. This indicates that the earthquake-triggered landslides in the frontal Longmenshan Mountains are a primary mechanism of long-term erosion.

2.6 Conclusion

There are far more scientific papers and reports about the Wenchuan-triggered landslides than are introduced in this paper; however, we have summarized the most relevant, important, and innovative aspects of the research. The 2008 Wenchuan earthquake triggered approximately 200 000 landslides, representing the largest number of landslides related to an individual earthquake ever recorded. It may be the result of a variety of factors, such as the ground-reaching source rupture, thrust faulting, strong ground shaking, steep terrain, low strength of slope materials, or other special geologic conditions. This event has also generated a very meaningful discussion on the relationship between seismic crust uplift and coseismic landslide denudation. Studies on the Wenchuan earthquake-triggered landslides will further our understanding of the occurrence and complex mechanisms of coseismic landslides, help in formulating landslide disaster prevention and mitigation strategies, and reveal their long-term impact on the landscape evolution of the affected area. In addition, they also provide important scientific reference for the research of other earthquakes. In brief, the Wenchuan earthquake event is a benchmark for the study of earthquake-triggered landslides.

Acknowledgements This study was supported by the National Natural Science Foundation of China (41661144037). We would like to thank Anthony Abram (http://www.uni-edit.net) for editing and proofreading this manuscript.

References

1. Chen Q, Cheng H, Yang Y, Liu G, Liu L (2014) Quantification of mass wasting volume associated with the giant landslide Daguangbao induced by the 2008 Wenchuan earthquake from persistent scatterer InSAR. Remote Sens Environ 152:125–135
2. Dai FC, Xu C, Yao X, Xu L, Tu XB, Gong QM (2011) Spatial distribution of landslides triggered by the 2008 Ms 8.0 Wenchuan earthquake, China. J Asian Earth Sci 40(4):883–895
3. de Michele M, Raucoules D, De Sigoyer J, Pubellier M, Chamot-Rooke N (2010) Three-dimensional surface displacement of the 2008 May 12 Sichuan earthquake (China) derived from synthetic aperture radar: evidence for rupture on a blind thrust. Geophys J Int 183(3):1097–1103
4. Ding H, Li Y, Ni S, Ma G, Shi Z, Zhao G, Yan L, Yan Z (2014) Increased sediment discharge driven by heavy rainfall after Wenchuan earthquake: a case study in the upper reaches of the Min River, Sichuan, China. Quatern Int 333:122–129
5. Godt J, Sener B, Verdin K, Wald D, Earle P, Harp E, Jibson R (2008) Rapid assessment of earthquake-induced landsliding. In: Proceedings of the first world landslide forum, United Nations University, Tokyo
6. Gorum T, Fan X, van Westen CJ, Huang R, Xu Q, Tang C, Wang G (2011) Distribution pattern of earthquake-induced landslides triggered by the 12 May 2008 Wenchuan earthquake. Geomorphology 133(3–4):152–167
7. Guo Z, Zhou C, Sun X, Zhang J (2010) The distribution of landslide triggered by Wenchuan earthquake and its causative factors. Earth Sci Front 17(5):234–242
8. Huang R, Li W (2009) Analysis of the geo-hazards triggered by the 12 May 2008 Wenchuan earthquake, China. Bull Eng Geol Env 68(3):363–371
9. Huang R, Li W (2009) Analysis on the number and density of landslides triggered by the 2008 Wenchuan earthquake, China. J Geol Hazards Environ Pres 20(3):1–7
10. Huang R, Fan X (2013) The landslide story. Nat Geosci 6(5):325–326
11. Huang R, Pei X, Li T (2008) Basic characteristics and formation mechanism of the largest scale landslide at Daguangbao occurred during the Wenchuan earthquake. J Eng Geol 16(6):730–741
12. Hubbard J, Shaw JH (2009) Uplift of the Longmen Shan and Tibetan plateau, and the 2008 Wenchuan (M = 7.9) earthquake. Nature 458(12):194–197
13. Li Y, Chen G, Tang C, Zhou G, Zheng L (2012) Rainfall and earthquake-induced landslide susceptibility assessment using GIS and artificial neural network. Nat Hazards Earth Syst Sci 12(8):2719–2729
14. Li W, Huang R, Tang C, Xu Q, van Westen CJ (2013) Co-seismic landslide inventory and susceptibility mapping in the 2008 Wenchuan earthquake disaster area, China. J Mountain Sci 10(3):339–354
15. Li G, West AJ, Densmore AL, Hammond DE, Jin Z, Zhang F, Wang J, Hilton RG (2016) Connectivity of earthquake-triggered landslides with the fluvial network: implications for landslide sediment transport after the 2008 Wenchuan earthquake. J Geophys Res Earth Surf 121(4):703–724
16. Li G, West AJ, Densmore AL, Jin Z, Parker RN, Hilton RG (2014) Seismic mountain building: landslides associated with the 2008 Wenchuan earthquake in the context of a generalized model for earthquake volume balance. Geochem Geophys Geosyst 15(4):833–844

17. Li G, West AJ, Densmore AL, Jin Z, Zhang F, Wang J, Clark M, Hilton RG (2017) Earthquakes drive focused denudation along a tectonically active mountain front. Earth Planet Sci Lett 472:253–265
18. Parker RN, Densmore AL, Rosser NJ, De Michele M, Li Y, Huang RQ, Whadcoat S, Petley DN (2011) Mass wasting triggered by the 2008 Wenchuan earthquake is greater than orogenic growth. Nat Geosci 4(7):449–452
19. Parsons T, Ji C, Kirby E (2008) Stress changes from the 2008 Wenchuan earthquake and increased hazard in the Sichuan basin. Nature 454(7203):509–510
20. Qi S, Xu Q, Lan H, Zhang B, Liu J (2010) Spatial distribution analysis of landslides triggered by 2008.5.12 Wenchuan earthquake, China. Eng Geol 116(1–2):95–108
21. Shen Z-K, Sun J, Zhang P, Wan Y, Wang M, Bürgmann R, Zeng Y, Gan W, Liao H, Wang Q (2009) Slip maxima at fault junctions and rupturing of barriers during the 2008 Wenchuan earthquake. Nat Geosci 2(10):718–724
22. Song Y, Gong J, Gao S, Wang D, Cui T, Li Y, Wei B (2012) Susceptibility assessment of earthquake-induced landslides using Bayesian network: a case study in Beichuan, China. Comput Geosci 42:189–199
23. Tang C, Van Westen CJ, Tanyas H, Jetten VG (2016) Analysing post-earthquake landslide activity using multi-temporal landslide inventories near the epicentral area of the 2008 Wenchuan earthquake. Nat Hazards Earth Syst Sci 16:2641–2655
24. Wang J, Jin Z, Hilton RG, Zhang F, Densmore AL, Li G, West AJ (2015) Controls on fluvial evacuation of sediment from earthquake-triggered landslides. Geology 43(2):115–118
25. Wang Q, Qiao XJ, Lan QG, Jeffrey F, Yang SM, Xu CJ, Yang YL, You XZ, Tan K, Chen G (2011) Rupture of deep faults in the 2008 Wenchuan earthquake and uplift of the Longmen Shan. Nat Geosci 4(9):634–640
26. Wang T, Wu S, Shi J, Xin P (2013) Case study on rapid assessment of regional seismic landslide hazard based on simplified Newmark displacement model: Wenchuan Ms8.0 earthquake. J Eng Geol 21(1):16–24
27. Wang Y, Song C, Lin Q, Li J (2016) Occurrence probability assessment of earthquake-triggered landslides with Newmark displacement values and logistic regression: the Wenchuan earthquake, China. Geomorphology 258:108–119
28. Xu C (2010) Spatial distribution and susceptibility mapping of the Wenchuan earthquake-triggered landslides. Institute of Geology and Geophysics, Chinese Academy of Sciences, Beijing
29. Xu C (2013) Actual landslides as most objective standard for validation of landslide hazard assessment result. J Eng Geol 21(6):908–911
30. Xu C (2015) Preparation of earthquake-triggered landslide inventory maps using remote sensing and GIS technologies: principles and case studies. Geosci Front 6(6):825–836
31. Xu C, Dai F, Chen J, Tu X, Xu L, Li W, Tian W, Cao Y, Yao X (2009) Identification and analysis of secondary geological hazards triggered by a magnitude 8.0 Wenchuan earthquake. J Remote Sens 13(4):745–762
32. Xu C, Dai F, Xu X (2010) Wenchuan earthquake induced landslides: an overview. Geol Rev 56(6):860–874
33. Xu C, Dai F, Xu X, Lee YH (2012) GIS-based support vector machine modeling of earthquake-triggered landslide susceptibility in the Jianjiang River watershed, China. Geomorphology 145–146:70–80
34. Xu C, Xu X (2012) Comment on "Spatial distribution analysis of landslides triggered by 2008.5.12 Wenchuan Earthquake, China" by Qi S, Xu Q, Lan H, Zhang B, Liu J [Engineering Geology 116 (2010) 95–108]. Eng Geol 133–134:40–42
35. Xu C, Xu X, Dai F, Saraf AK (2012) Comparison of different models for susceptibility mapping of earthquake triggered landslides related with the 2008 Wenchuan earthquake in China. Comput Geosci 46:317–329
36. Xu C, Xu X, Dai F, Xiao J, Tan X, Yuan R (2012) Landslide hazard mapping using GIS and weight of evidence model in Qingshui River watershed of 2008 Wenchuan earthquake struck region. J Earth Sci 23(1):97–120

37. Xu C, Xu X, Gorum T, van Westen CJ, Fan X (2014) Did the 2008 Wenchuan earthquake lead to a net volume loss? Landslide science for a safer geoenvironment. In: Proceedings of world landslide forum 3, 2–6 June 2014, Beijing, vol 3, pp 19–196
38. Xu C, Xu X, Shen L, Yao Q, Tan X, Kang W, Ma S, Wu X, Cai J, Gao M, Li K (2016) Optimized volume models of earthquake-triggered landslides. Sci Rep 6:29797
39. Xu C, Xu X, Yao Q, Wang Y (2013) GIS-based bivariate statistical modelling for earthquake-triggered landslides susceptibility mapping related to the 2008 Wenchuan earthquake, China. Q J Eng Geol Hydrogeol 46(2):221–236
40. Xu C, Xu X, Yao X, Dai F (2014) Three (nearly) complete inventories of landslides triggered by the May 12, 2008 Wenchuan Mw 7.9 earthquake of China and their spatial distribution statistical analysis. Landslides 11(3):441–461
41. Xu W-J, Xu Q, Wang Y-J (2013) The mechanism of high-speed motion and damming of the Tangjiashan landslide. Eng Geol 157:8–20
42. Xu X, Wen X, Yu G, Chen G, Klinger Y, Hubbard J, Shaw J (2009) Coseismic reverse-and oblique-slip surface faulting generated by the 2008 Mw 7.9 Wenchuan earthquake, China. Geology 37(6):515–518
43. Yin Y (2008) Research on the geo-hazards triggered by Wenchuan earthquake, Sichuan. J Eng Geol 16(4):433–444
44. Yin Y, Li B, Wang W (2015) Dynamic analysis of the stabilized Wangjiayan landslide in the Wenchuan Ms 8.0 earthquake and aftershocks. Landslides 12(3):537–547
45. Yin Y, Wang F, Sun P (2009) Landslide hazards triggered by the 2008 Wenchuan earthquake, Sichuan. China. Landslides 6(2):139–152
46. Yin YP, Cheng YL, Wang J, Wang M, Liu B, Song Y, Liang JT (2011) Remote sensing research on Daguangbao gigantic rock-slide triggered by Wenchuan earthquake. J Eng Geol 19(5):674–684

Chapter 3
Late-Cenozoic Faulting History of the Central Longmenshan Thrust Belt and Its Tectonic Implications

Xi-Bin Tan, Yuan-Hsi Lee, Xi-Wei Xu, Yi-Duo Liu, and Ren-Qi Lu

Abstract The central segment of the Longmenshan thrust belt (LTB) has attracted the attention of many researchers because of the devastating 2008 Wenchuan earthquake and the controversial continental tectonic mechanisms of the LTB. In this study, we integrate low-temperature thermochronology published data to characterize the late-Cenozoic exhumation and faulting history of the central LTB. The results reveal that the three major faults in the study area have been accommodating reverse faulting since the middle Miocene and that the initial timing of these faults suggests basinward in-sequence propagation of the deformation front. The late-Cenozoic fault activity suggests that upper crustal thrusting and shortening is one of the major mechanisms for the LTB's uplift. Furthermore, middle-lower crustal deformation cannot be neglected when considering the orogenesis along the plateau margin. Moreover, comparisons of co-seismic faulting, co-seismic landslide, fluvial intensity, and rock abrasion data indicates that the along-strike variation of fluvial incision intensity throughout the central LTB is a fundamental driver for the along-strike variation of the topography and fault activity in the LTB.

X.-B. Tan (✉) · X.-W. Xu · R.-Q. Lu
Key Laboratory of Active Tectonics and Volcano, Institute of Geology, China Earthquake Administration, Beijing, China
e-mail: tanxibin@163.com

Y.-D. Liu
Department of Earth and Atmospheric Sciences, University of Houston, Houston, Texas, USA

Y.-H. Lee
Department of Earth and Environmental Sciences, National Chung-Cheng University, Chia-Yi, Chiayi, Taiwan

X.-W. Xu
Institute of Crustal Dynamics, China Earthquake Administration, Beijing, China

C.-H. Lo et al. (eds.), *Earthquake Geology and Tectonophysics around Eastern Tibet and Taiwan*, Atmosphere, Earth, Ocean & Space,
https://doi.org/10.1007/978-981-15-6210-5_3

3.1 Introduction

The Cenozoic India–Eurasia collision has profoundly influenced the tectonics, landscape, and climate in Asia [29, 44]. The primary product of the collision was the Tibetan Plateau, which now serves as a major testing ground for competing models based on the tectonic mechanisms of the continental crust and the lithosphere [7, 12, 33, 45].

Along the eastern Tibetan Plateau margin, the eastward-moving Bayan Har block is blocked by the strong South China block, forming the Longmen Shan ("Shan" means mountains in Chinese) that is ~4000 m higher than the neighboring Sichuan Basin (Fig. 3.1). The Longmenshan thrust belt (LTB) is comprised of three NW-dipping major faults, described from west to east as, the Wenchuan–Maoxiao fault

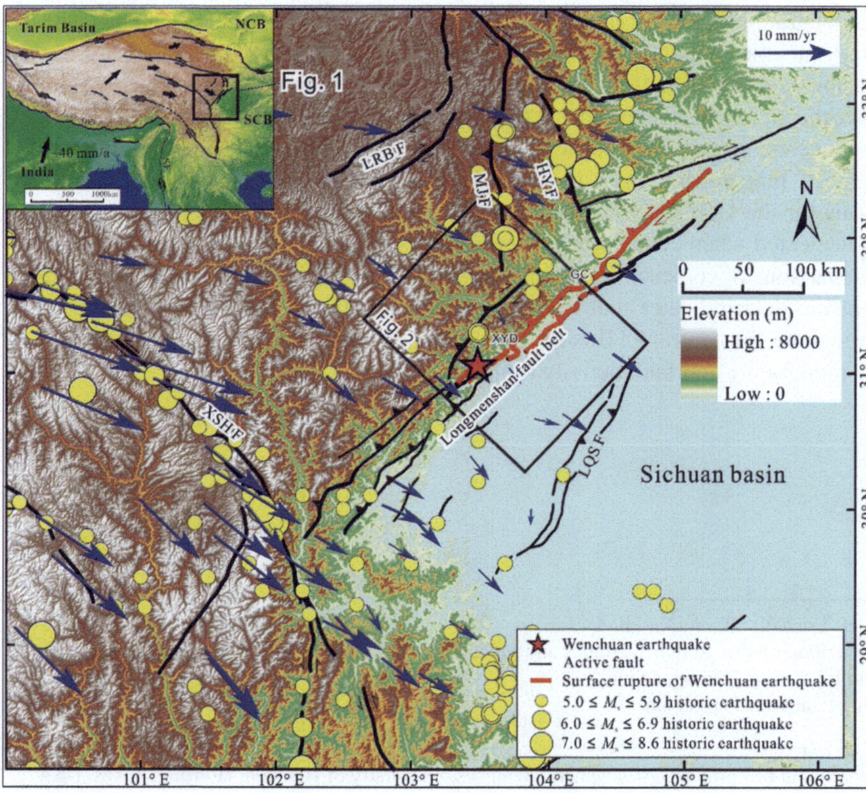

Fig. 3.1 Topographic map and neotectonic setting of the eastern Tibetan Plateau margin. Historical earthquake distribution and the surface rupture of the Wenchuan earthquake are revised from [64]. GPS vectors (blue arrows) from Liang et al. [25]. Abbreviations: LQSF: Longquanshan fault; LRBF: Longriba fault; MJF: Mingjiang fault; XSHF: Xianshuihe fault; HYF: Huya fault; XYD: Xiaoyudong step-over; GC: Gaochuan step-over; NCB: North China block; SCB: South China block

(WMF), the Beichuan–Yingxiu fault (BYF), and the Jiangyou–Guanxian fault (JGF) (Fig. 3.1 and 3.2). Significant shortening was not detected across the LTB by GPS measurements at a timescale of ~10 years [5] until the catastrophic 12 May, 2008 event where a Mw7.9 Wenchuan earthquake impacted the LTB. This earthquake caused active shortening across the LTB, which has since become the focus of numerous studies [4, 19, 58, 62]. Researchers recommenced the debate on the active mechanism responsible for the uplift of the Longmen Shan based on surveying co-seismic surface ruptures and conducting other new geophysical and geochronology

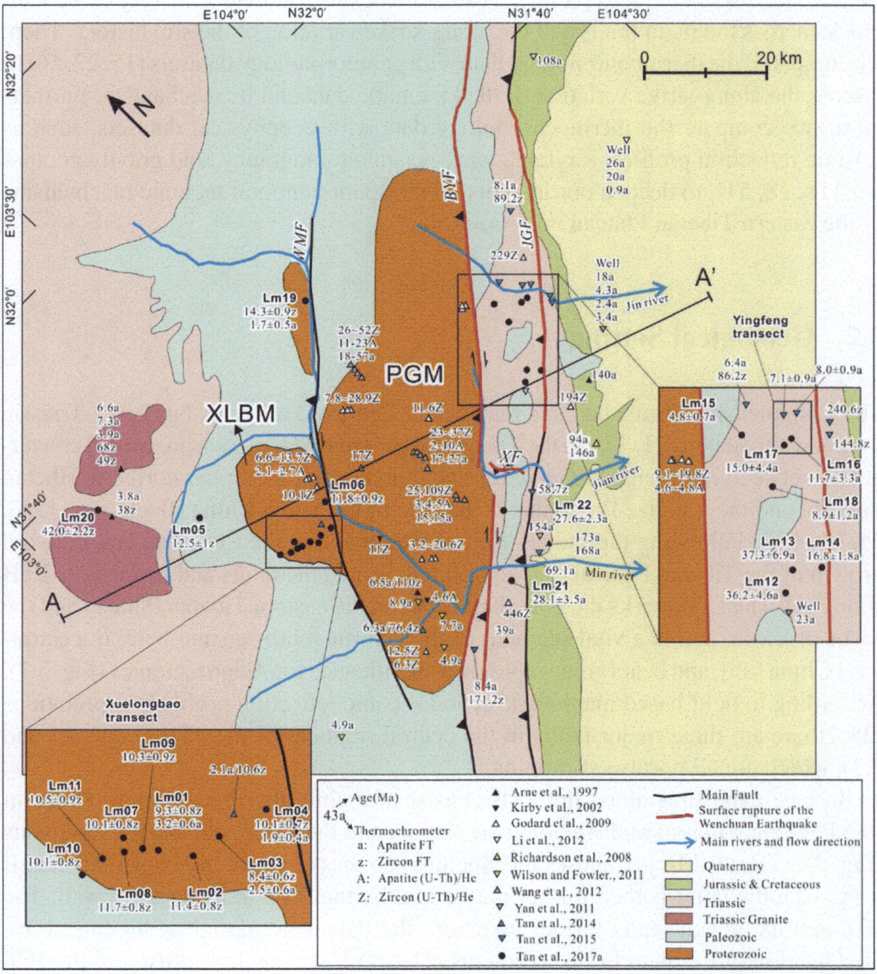

Fig. 3.2 Geological map of the central segment of the LTB including low-temperature thermochronology data. Revised from Tan et al. [40]. The location shown in the black box is depicted in Fig. 3.1. Abbreviations: XLBM: Xuelongbao Massif; PGM: Pengguan Massif, BYF: Beichuan–Yingxiu fault; JGF: Jiangyou–Guanxian fault; WMF: Wenchuan–Maoxiao fault

studies. Two sets of viewpoints have been formed, which suggest that the uplift was either caused by upper crustal shortening [19, 35, 58] or lower crustal thickening [4, 67]; Zhang et al. [65].

Low-temperature thermochronology dating is an ideal method for documenting the uplift–exhumation history of rocks in fault systems and has been extensively applied in the LTB [14, 40, 47]. Such information provides critical constraints for evaluating the above-mentioned controversies and for understanding the growth mechanism of the eastern Tibetan Plateau [22, 24, 39]. In this study, we compiled low-temperature thermochronology data in the central LTB from eleven studies [1, 14, 22, 31, 38–40, 47, 54, 61] (Fig. 3.2), to evaluate the exhumation history of the fault blocks across the main faults and the along-strike variation of the slip history. Then, we compared the thermochronology data with geomorphology datasets [15, 22, 56] to discuss the along-strike variation of the kinematic data and its mechanism. Furthermore, we compare the thermochronology data with geophysical datasets, such as seismic reflection profiles, Rayleigh wave seismic tomography, and gravity anomalies [18, 28, 51], to deepen our insights on the spatiotemporal tectonic mechanisms of the eastern Tibetan Plateau margin.

3.2 Geological Setting

The Longmenshan thrust belt is a reactivated Cenozoic orogeny, built on a Triassic fold-and-thrust belt [21, 37]. In the Triassic, it accommodated transpressional convergence between the Songpan-Garze terrane, a major terrane in the northern Tibetan Plateau, and the Yangtze Block that is the core of the South China Block (e.g., [3, 8, 55, 61, 63]). Following the Indian–Eurasian collision in the Cenozoic era, outward growth of the Tibetan Plateau [33, 45, 63] reactivated the faults and formed the LTB (Fig. 3.1). The LTB marks the boundary between the eastern Tibetan Plateau and the Sichuan Basin, and is a vital segment of the "North-south seismic belt" in central-west China [49], and bears great geological significance for Asian tectonics (Fig. 3.1). According to field-based mapping [58] and seismic reflection profile interpretations [28], there are three major faults in the central segment of the LTB (Figs. 3.1 and 3.2), which are all northwest-dipping.

Surface exposures along the LTB consist of mainly the Precambrian basement (the Pengguan massif and Xuelongbao massif) and the Paleozoic marine sediments (Fig. 3.2). The WMF juxtaposes the Xuelongbao massif above the Pengguan massif (Fig. 3.2); the latter is then further juxtaposed by the BYF at its hanging wall. The JGF and its branch faults further juxtapose the Paleozoic marine sediments above the Quaternary foreland basin sediments. The BYF is a dextral thrust and the JGF shows pure thrust, based on the faulting behaviors arising from the 2008 Wenchuan earthquake that caused ~ 240 km and ~ 70 km long surface ruptures along the two fault lines, respectively (Fig. 3.1) [58]. However, regarding the WMF, while the dextral movement is unequivocal [3, 9], there is debate on whether the mechanism of its late-Cenozoic dip-slip is top-to-SE thrusting or top-to-NW normal faulting [3, 46, 58].

The deep structure under the LTB is another key in elucidating the deformation mechanism of the Tibetan Plateau (Fig. 3.2b). Several groups have reported the Moho's distribution, fault geometry, and various crustal attributes using different methods, and most of the previous studies agree upon an increase of the Moho depth below the LTB from the Sichuan Basin to the Tibetan Plateau, and that a relatively weak middle-low crust occurs under eastern Tibet [26, 28, 32, 48, 50, 66, 67].

Upper crustal shortening along the LTB has been studied at both long-term (Ma) [19, 20] and short-term timescales (Ka) [5, 35, 58]. A recent analysis of seismic reflection profiles and surface geology from the LTB to the Sichuan Basin shows over 45 km, or 47%, of upper crustal shortening over a distance of ~50 km across the LTB, implying the significance of upper crustal shortening in building up the high topography of the LTB [19, 20]. On a short-term scale, inter-seismic GPS observations have demonstrated that the upper crustal NW-SE shortening across the LTB is ~1–3 mm/yr [5]. In addition, field surveys [27, 58] and measured differences in the positions of geometrical markers, observed from satellite images before and after the Wenchuan earthquake [35], show that the co-seismic NW-SE shortening by the Wenchuan earthquake was ~6–10 m.

3.3 Exhumation History of Different Blocks

The three major faults (WMF, BYF, and JGF) divide the central Longmen Shan into four blocks (Fig. 3.2). For the Xuelongbao massif, positioned in the hanging-wall block of the WMF, a number of thermochronological datasets characterize the occurrence of a prominent late-Cenozoic cooling history [14, 40]. The Pengguan massif occupies the majority of the area situated between the WMF and the BYF (Fig. 3.2). Several groups have published large amounts of thermochronology data from the Pengguan massif to obtain the exhumation history [14, 47]. Between the BYF and the JGF, approximately twenty samples were used for AFT dating to reveal the cooling history and along-strike variation of faulting activity along the JGF [39, 40]. To the east of the JGF is mainly the Chengdu plain, with little thermochronology data available, most of which is older than 100 Ma. In this chapter, and based on the published data, we summarize the exhumation or cooling history of the Xuelongbao massif, the Pengguan massif, and the range frontal region between the BYF and JGF, respectively.

3.3.1 Xuelongbao Massif

The Xuelongbao massif is located at the hanging-wall block of the WMF, occupying the highest position structurally. It also hosts the highest peak (elevation of

5285 m, measured from Google Earth) in the central LTB. Exhumation history of the Xuelongbao massif provides a critical control when delineating the WMF's faulting activity. The thermochronology data presented (Fig. 3.3a, b) is mainly derived from [14] and Tan et al. [40]. Four low-temperature thermochronometers (apatite fission track, AFT; zircon fission track, ZFT; apatite (U-Th)/He, AHe; and zircon (U-Th)/He, ZHe) all yield mid-Miocene–Pleistocene cooling ages from the immediate hanging wall of the WMF. Farther west, granitoid plutons that were emplaced in the Triassic yield Paleocene–Eocene ZFT ages and late Miocene to Pliocene AFT ages. Although the elevation difference of the profile is not large enough to show a clear inflection of the exhumation rate, average exhumation rates for the AFT and ZFT of ~0.30 mm/yr at 1.5–2.3 Ma and ~1.45 mm/yr at 10–11 Ma, respectively, can be derived through the linear fitting of the AFT and ZFT ages against elevation (Fig. 3.3a). Based on

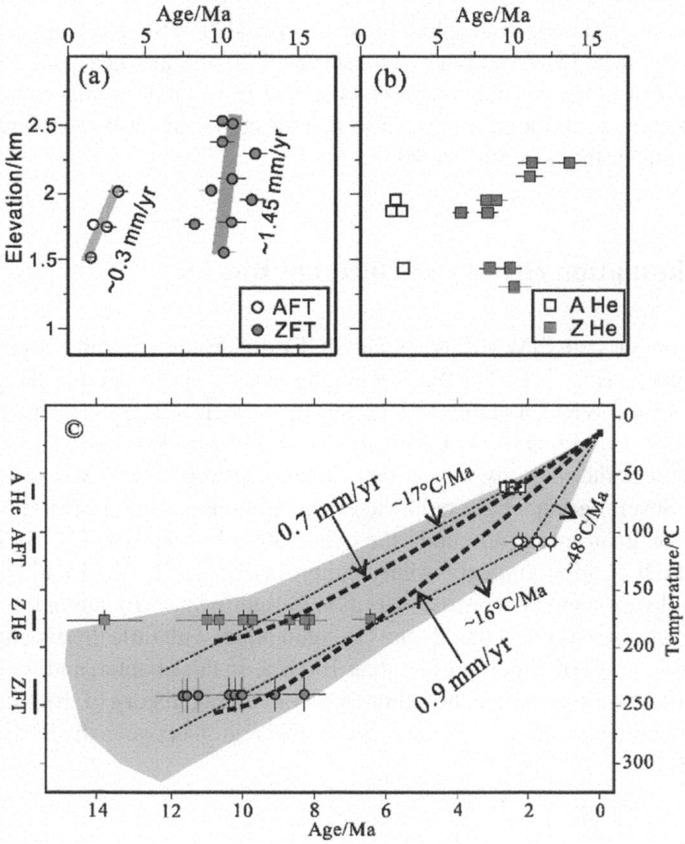

Fig. 3.3 Xuelongbao transects age-elevation plots and exhumation history, from Tan et al. [40]. **a** Fission track age-elevation plots for the Xuelongbao transect. **b** (U-Th)/He age-elevation plots at the Xuelongbao Massif, revised from [14]. **c** Cooling history of the Xuelongbao Massif based on fission track and (U-Th)/He data

the different annealing/closure temperatures of these thermochronometers, Fig. 3.3c shows a series of cooling ages for the two different vertical profiles.

To further quantify the cooling history, we used the TERRA software [11] to model the cooling rate since 11 Ma under various rates of steady-state erosion, with the following boundary conditions: a surface temperature of 15 °C, a basal temperature gradient of 25 °C/km, a maximum model depth of 50 km, and diffusivity of 1.3636 $\times$ 10^{-6} m^2/s. The basal temperature gradient lies between the average geothermal gradient of ~27 °C/km in the Songpan-Aba region [6] and that of ~22.8 °C/km in the Sichuan basin [57]. The results show that the observed cooling rates best match the data when the erosion rates are ~0.7 mm/yr and ~0.9 mm/yr for the north and south transects, respectively (Fig. 3.3c). The minor difference between the two transects may be caused by sampling bias, analytical errors, or, along-strike variations in lithology, deformation style, and/or exhumation; however, in general, a cooling rate of 0.7–0.9 mm/yr characterizes the exhumation history of the Xuelongbao massif.

3.3.2 Pengguan Massif

The Pengguan massif occupies the backbone of the central LTB, and several research groups have obtained a number of low-temperature thermochronology datasets for the massif [1, 14, 22, 38, 47, 54]. Overall, the ZFT and ZHe ages range widely from the mid-Cretaceous to the Miocene, while the AFT and AHe ages are predominantly of the Miocene–Pleistocene (Fig. 3.2). Kirby et al. [22] modeled the temperature–time paths from inverse modeling of the kinetic parameters and domain structure on the 40Ar/39Ar release spectrum and acquired the thermal history by combining the ZFT and AFT ages (Fig. 3.4). The thermal history suggests that the sample underwent extremely slow cooling during the Cretaceous and early Tertiary, with a dramatic increase in cooling rates (~20–50 °C/m.y.) at ca. 11–12 Ma (red line in the Fig. 3.4). Godard et al. [14] reported (U-Th)/He ages from several profiles

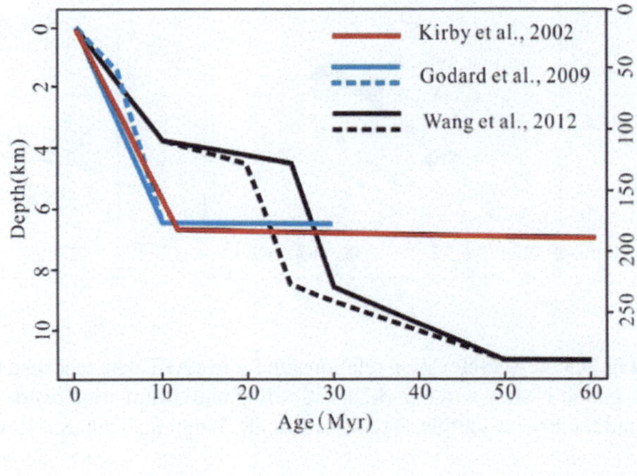

Fig. 3.4 Exhumation history of the Pengguan massif. Revised from [14, 47]

in the Pengguan massif and modeled the cooling history of the Pengguan massif. The exhumation history, shown by the blue lines in Fig. 3.4, is similar to Kirby et al. [22]'s cooling history. Wang et al. [47] analyzed the fission track and (U-Th)/He thermochronology from a profile with a 3 km difference in elevation in the Pengguan massif. Their exhumation modeling shows a two-phase cooling history (black lines in Fig. 3.4): comprised of a younger stage spanning from 10 to 15 Ma to present that is consistent with previous studies (with a lower exhumation rate and smaller exhumation amount), and an earlier stage between 30 and 25 Ma. Despite the discordance in the cooling histories during the early Cenozoic, the late-Cenozoic cooling of the Pengguan massif occurring since ~10 Ma is clear, with a 4–6.5 km exhumation (Fig. 3.4).

3.3.3 *Range Frontal Area Between the Beichuan–Yingxiu and the Jiangyou–Guangxian Fault*

The thermochronology data recorded from the Longmen Shan range frontal area between the BYF and the JGF is primarily taken from [1, 39, 40] (Fig. 3.2). Tan et al. [40] obtained the cooling history of a profile and two individual samples. In general, ZFT and ZHe dating yielded only pre-Cenozoic cooling ages, whereas AFT ages in this area were of the Eocene to mid-Miocene (Fig. 3.5).

Four samples from the Yingfeng transect illustrate a linear relationship in the age-elevation diagram, showing a slow exhumation between 7 and 15 Ma (Fig. 3.5a). Figure 3.5b shows the exhumation histories of the Lm21, Lm22, and the Yingfeng transect [40]. The cooling rate of the Yingfeng transect from 15 Ma to present was calculated under reasonable assumptions, including a surface temperature of 15 °C, a constant thermal gradient of 25 °C/km, and an annealing temperature of 110 ± 10 °C [2, 10]. The cooling history from 35 to 15 Ma was estimated using samples Lm21 and

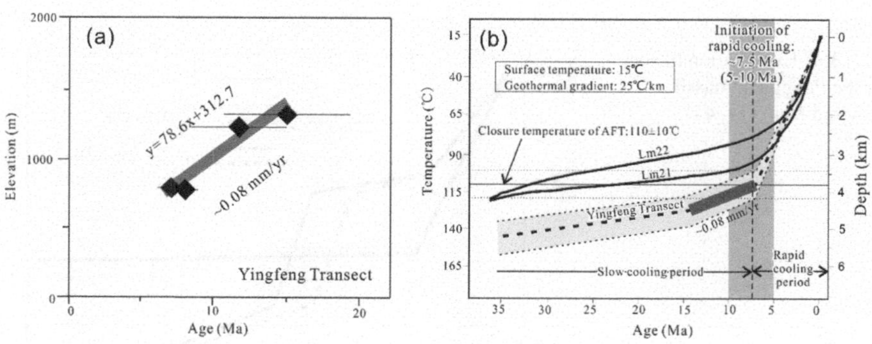

Fig. 3.5 **a** Age-elevation relationship for the AFT data recorded from the Yingfeng transect, whose location is shown in Fig. 3.2a. **b** Cooling and exhumation history of the Lm21 and Lm22 samples, and the lowest sample obtained from the Yingfeng Transect. Revised from Tan et al. [40, 42]

Lm22. All models (Fig. 3.5b) show rapid cooling from ~7.5 Ma to present, with a total temperature drop of 65–105 °C, while the cooling rate before ~7.5 Ma is less than 1 °C/Ma. The total exhumation amount was ~2.5–4.5 km given an average geothermal gradient of 25 °C/km, which is a little higher than the 22.8 °C/km recorded in the Sichuan basin [57].

3.4 Discussion

3.4.1 Late-Cenozoic Activity on the Major Faults

Analysis of the exhumation histories of the Xuelongbao massif and the range frontal area acquired by Tan et al. [40] along with those of the Pengguan massif, acquired through previous studies [14, 47] show that the major faults in central LTB have been accommodating shortening and uplifting since the middle Miocene. The results additionally verify the reverse faulting activity on the WMF, BYF, and JGF (Fig. 3.6), consistent with the faulting activity occurring in the southern segment of the LTB [38, 46]. The exhumation processes of the different regions also suggest basinward in-sequence propagation.

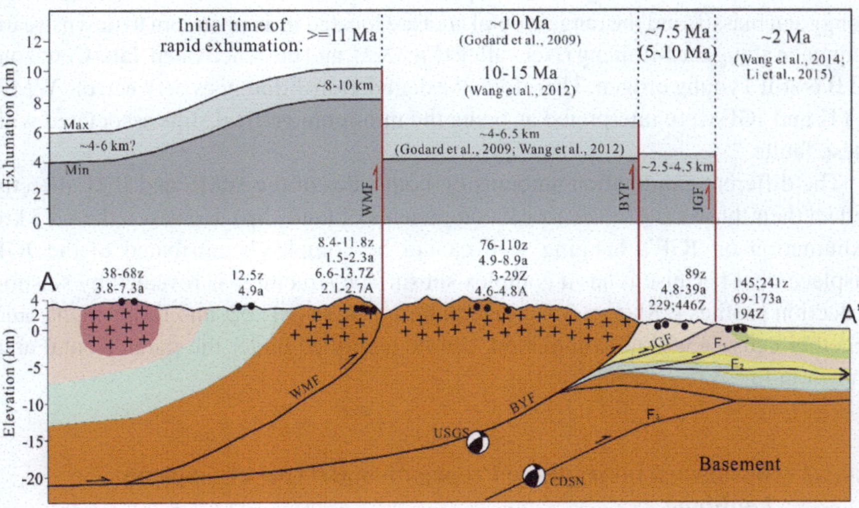

Fig. 3.6 Deep structure of the LTB central segment, revised from Lu et al. [28] and low-temperature thermochronology data, from Tan et al. [40]. Location of the line A–A′ is shown in Fig. 3.2. Low-temperature thermochronological ages are shown for samples taken from the bottom of the valley. The legend for stratum and thermochronometer is as per Fig. 3.2

3.4.1.1 Dip-Slip Along the Three Major Faults

The amount of late-Cenozoic total denudation observed decreases eastward across the three faults. Denudation of the Xuelongbao massif from ~11 Ma to present was ~8–10 km (average ~9 km) (Fig. 3.3). For the Pengguan massif, it was ~6.5 km [14] or ~4 km [47] (Fig. 3.4), and for the range frontal area (hanging-wall block of the JGF), it was 2.5–4.5 km (an average of ~3.5 km) (Fig. 3.5). Keeping the difference between rock uplift and denudation in mind [13], we determined the amount of denudation in terms of the amount of rock uplift, assuming that uplift and erosion are balanced [15]. Under the assumption that at the local scale, similar elements of the landscape on both sides of a fault are equilibrated, and that some estimation of the vertical throw rates for such faults can be inferred from the difference in erosion rates between the two compartments [15], the difference in the amount of denudation on both sides of the WMF, BYF, and JGF (including the blind faults to the east) are ~2.5–5 km, ~0.5–3 km, and ~3.5 km, respectively. The pattern of differential denudation across each fault indicates that since the middle Miocene, the three major faults in the central LTB have all been reverse faults.

The WMF, BYF, and JGF with the blind thrust faults have accommodated at least ~2.5–5 km, ~0.5–3 km, and ~3.5 km of vertical slip, respectively. In an active orogen, the difference between uplift and erosion is usually greater in the upstream region than in the downstream. Particularly, the amount of erosion is small when compared to the uplift that occurs during the initial phase of orogeny. The Xuelongbao massif, the Pengguan massif, and the range frontal area are located in an upstream to downstream sequence along the Minjiang river valley (Fig. 3.2), and the reactivated, late-Cenozoic LTB is still a young orogen. Therefore, the denudation differences between the WMF, BYF, and JGF were interpreted as being the minimum vertical slips associated with these faults.

The different exhumation amounts on both sides of the WMF and BYF directly reflect their throws (i.e., the vertical components of fault slip). However, the ~3.5 km exhumation on JGF's hanging wall cannot be completely attributed to the JGF displacement because it has a complex subsurface structure, as revealed by seismic reflection profiles [28] (Fig. 3.6). Several thrust faults (F1, F2, and F3), which could together cause ~3.5 km exhumation, can be identified under the range frontal area shown in Fig. 3.6.

3.4.1.2 Basinward in-Sequence Propagation and Out-of-Sequence Faulting

Exhumation histories of the Xuelongbao massif, the Pengguan massif, and the range frontal area constrain the initial timing of the main faults in the central LTB. The Xuelongbao massif began rapid exhumation after ~11 Ma, and the onset of rapid exhumation of the Pengguan massif was either at ~10 Ma [14] or 10–15 Ma [47] (following an older cooling event at ca. 30–25 Ma). The onset of rapid exhumation in the range frontal area was recorded at ~ 7.5 (5–10) Ma, according to fission track length modeling and age-elevation linear fitting, performed on profile samples (Fig. 3.5b) [40]. The blind faults to the east of the JGF in the Sichuan basin became active at ~2 Ma [23, 48, 50]. Although the exact time of initiation of the rapid exhumation of the Xuelongbao massif is not well understood; however, present results (Fig. 3.6) suggest basinward in-sequence propagation of the central LTB major faults occurred during the mid-Miocene to Pliocene, which has also been identified in the south LTB [38].

Out-of-sequence deformation in the central LTB emerged during the Quaternary and possibly during an earlier epoch. The Xuelongbao massif underwent rapid exhumation again, which is based on the 2–3 Ma AFT ages that are noticeably younger than those of the Pengguan massif (~ 5–9 Ma). This indicates the occurrence of Quaternary reverse faulting activity on the WMF. Therefore, activities of the range frontal faults (i.e., the JGF and other blind faults) did not lead to inactivity in the hinterland fault; instead, it is suggested that it is attributed to out-of-sequence thrusting of the LTB [27, 30, 46]. Sandbox analog modeling further supports this notion [36]. A number of factors may have resulted in the out-of-sequence thrusting in the LTB, including strain-hardening on the BYF and JGF due to the Yangtze Craton underthrusting, focused fluvial and glacial erosion in the Xuelongbao massif following the onset of the Asian monsoon, reactivated detachment caused by newly added partial melt pockets in the middle crust, and preexisting mechanical heterogeneities present in the crust. Further work is needed to elucidate the likely mechanism and assess the seismic hazard risk.

3.4.2 Along-Strike Variation and Its Mechanism

The AFT data further reveals the along-strike variation upon JGF's hanging wall, where the AFT ages gradually increased from northeast to southwest, except for two samples (Lm12 and Lm13) with AFT ages of >30 Ma, which may have been influenced by their relatively high elevations (Fig. 3.7f).

Recent sandbox modeling of evolving thrust wedges reveals the effects of preexisting topographic relief on thrust faults [36, 42]. Figure 3.7c shows the topographic difference along the BYF's strike, which shows that the northeastern segment (Longmen sub-segment, abbreviate to LMSS) of the Pengguan massif has a higher topographic relief than its southwestern segment (Hongkou sub-segment, abbreviate to HKSS), with the segments roughly separated by the Xiaoyudong fault. The sandbox modeling [36] reveals that the preexisting higher topographic relief caused stronger deformation at the footwall of the major fault (F1 in the paper by Sun et al. [36]), and the topographic relief at the BYF's footwall indeed showed higher topographic relief to the northeast of the Xiaoyudong (Fig. 3.7d). The along-strike variation in the thermochronology data is consistent with the sandbox modeling results presented by Sun et al. [36], along with the seismic reflection profile interpretations (Fig. 3.8) [28], and the co-seismic vertical slip that occurred along the JGF during the 2008 Wenchuan earthquake (Fig. 3.7e).

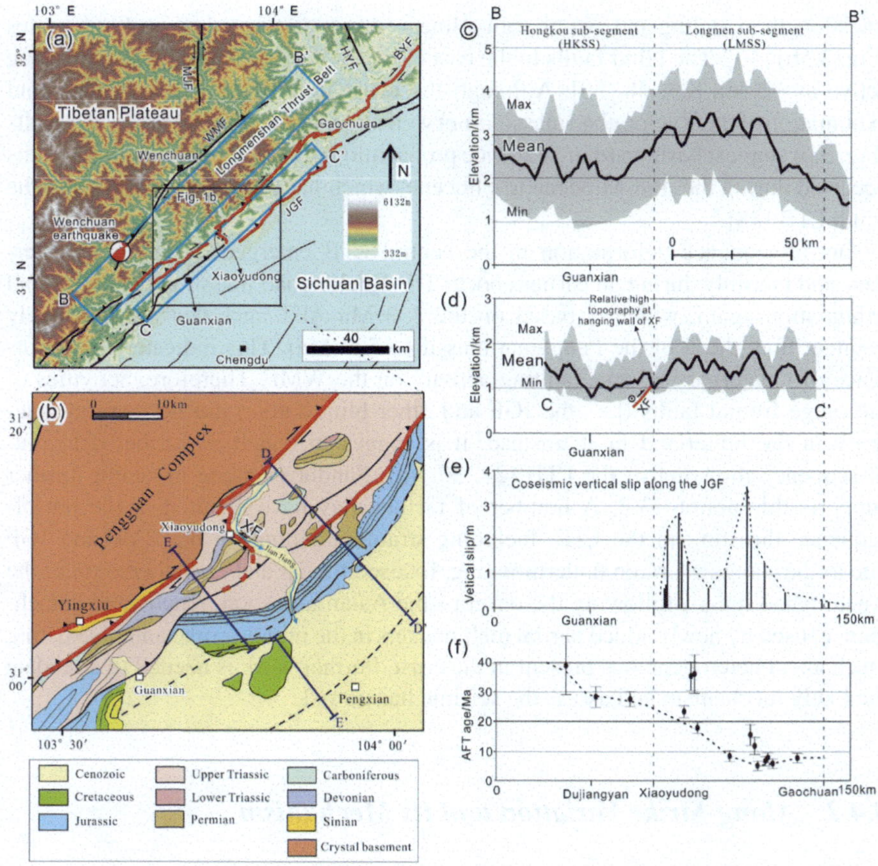

Fig. 3.7 **a** Topographic map of the research area showing active faults (black lines) and surface rupture from the 2008 Wenchuan earthquake (red lines). **b** Local geological map showing the position of seismic reflection profiles and co-seismic surface ruptures. Topographic swaths showing elevation features (mean, minimum, and maximum elevations) along the B–B′ and C–C′ profiles, respectively, revised from Lu et al. [28, 40, 42]. **c–d** The B–B′ and C–C′ profile locations are shown by the blue boxes presented in Fig. 3.1a. **e** Co-seismic vertical displacement along the JGF during the 2008 Wenchuan earthquake, revised from [64]. **f** The AFT age distributions present at the hanging wall and along the JGF strike. Abbreviations: MJF: Minjiang fault; BYF: Beichuan–Yingxiu fault; JGF: Jiangyou–Guanxian fault; WMF: Wenchuan–Maoxian fault

Seismic reflection profiles and thermochronology data revealed the long-term along-strike variation on the JGF's hanging wall, indicative of an increasing dip-slip component along the JGF located to the northeast of Xiaoyudong. Based on the sandbox modeling, it was found that preexisting topographic relief had caused the along-strike variation [36, 42]. A question is thus raised: how was the along-strike variation of the topography formed?

The topographical landscape is formed by the interaction between internal tectonic processes and the external denudation processes [15, 52, 53]. Therefore, two possible

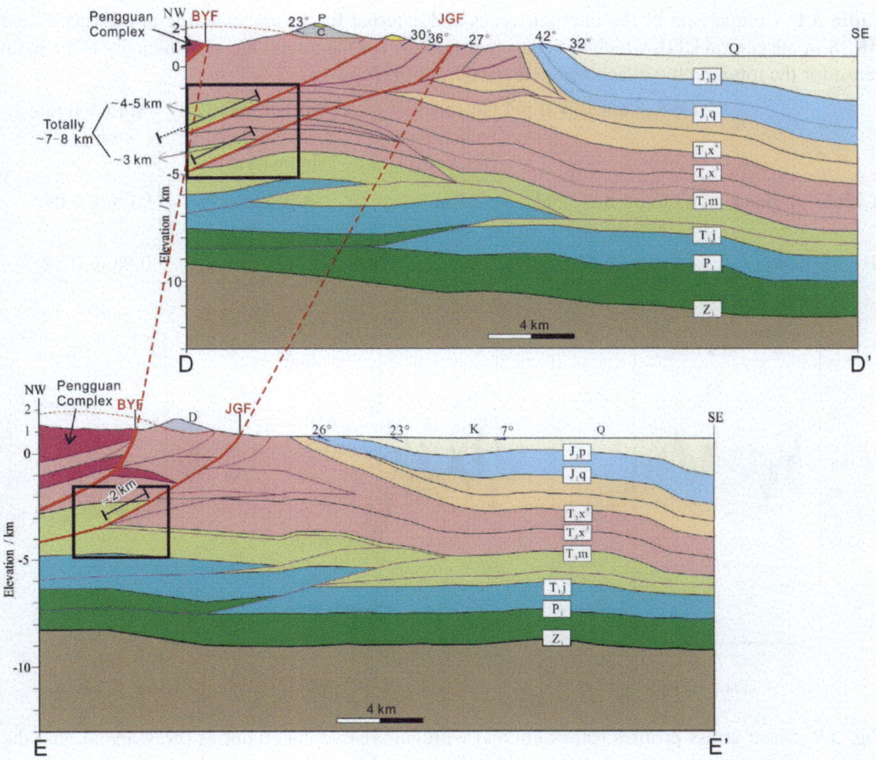

Fig. 3.8 Comparison of the interpreted structure of the central LTB from section C–C′ and D–D′. The profile locations are as per Fig. 3.7a. The black boxes show the dip-slip along the JGF and its branch, according to the Triassic sedimentary layer shown by a laurel-green color. The dip-slip on the JGF and its branch is approximately 7–8 km and 2–3 km for the C–C′ and D–D′ profile, respectively. Revised from Lu et al. [28] and Tan et al. [42]

scenarios exist for determining the mechanism producing the along-strike variation of the topography: (1) that thrust faulting in the LMSS is stronger than that in the HKSS, or (2) that erosion processes are weaker in the LMSS than the HKSS. The former mechanism violates the geological observations made along profile AA', as both the co-seismic thrust slip and long-term denudation in the LMSS are no greater than those in the HKSS [14, 34, 47, 58]. The first mechanism can thus be eliminated. According to the comparison between the relevant forces occurring in the LMSS and HKSS presented in Table 3.1, the second mechanism can readily explain the topographical observations [43]. Along-strike variation of the topography has resulted from different fluvial erosion intensities in the LMSS and HKSS, when the effects of precipitation, lithology (erodibility), and co-seismic landslides are similar between the two sub-segments (Figs. 3.9 and 3.10 and Table 3.1). Therefore, fluvial erosional processes, or external forces, appear to control the observed topographic

Table 3.1 Comparison of the internal forces and external forces associated with the LMSS and HKSS in the central LTB, which indicates that variation of the fluvial incision intensity is the main cause for the topographic differences. Revised from [43]

	Co-seismic dip-slip	Co-seismic landslides (thickness) (m)	Fluvial shear stress (incision intensity) (Pa)	Abrasion rate (%/km)
LMSS (Higher)	~3 m on BYF and ~1.5 m on JGF	~2.7	~150	0.08 ± 0.08
HKSS(Lower)	~5 m on BYF	~2.7	~400	0.08 ± 0.08

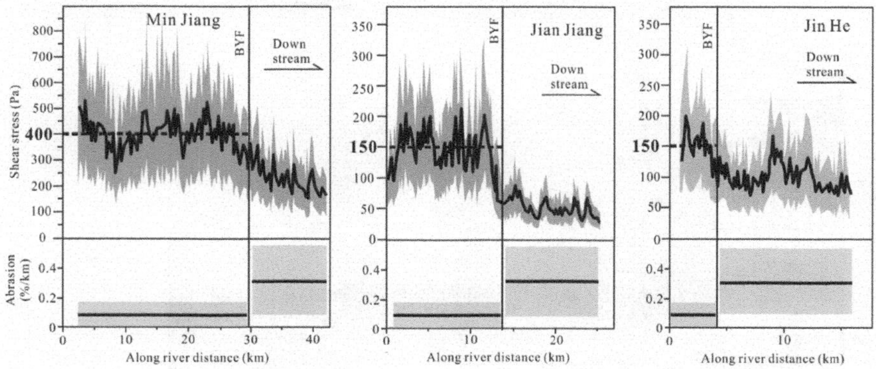

Fig. 3.9 Shear stress profiles (black curves) with cumulative uncertainties (gray areas), and the abrasion along the rivers (revised from [15]). Note the different scales of the vertical axes (shear stress, in Pascal) between Min Jiang and the other two rivers. See the river locations in Fig. 3.10

variations, which further played a critical role in defining the stress state and behaviors of the fault system in the central LTB [43].

3.4.3 *Evaluating Tectonic Models for the LTB*

It is generally regarded that the nucleation of earthquakes, thickened crust, and high topography typical of the LTB were responses to the Tibetan Plateau's ongoing outward growth [46]. However, the debate continues on whether the main mechanism of the LTB's uplift was due to upper crustal thrusting and shortening [19, 35, 45, 58] (Fig. 3.11a) or middle-lower crustal thickening and flow (e.g., [4, 65]) (Fig. 3.11b). One major difference in the predictions derived from the two models is the determination that the WMF was a reverse fault in the former model and a normal fault in the latter (Fig. 3.11a, b). The upper crustal thrusting model further predicts younger low-temperature thermochronology ages in the hanging wall of the WMF than in the footwall. In contrast, the footwall block should yield younger ages if normal faulting is applied in the middle- to lower crustal channel flow model.

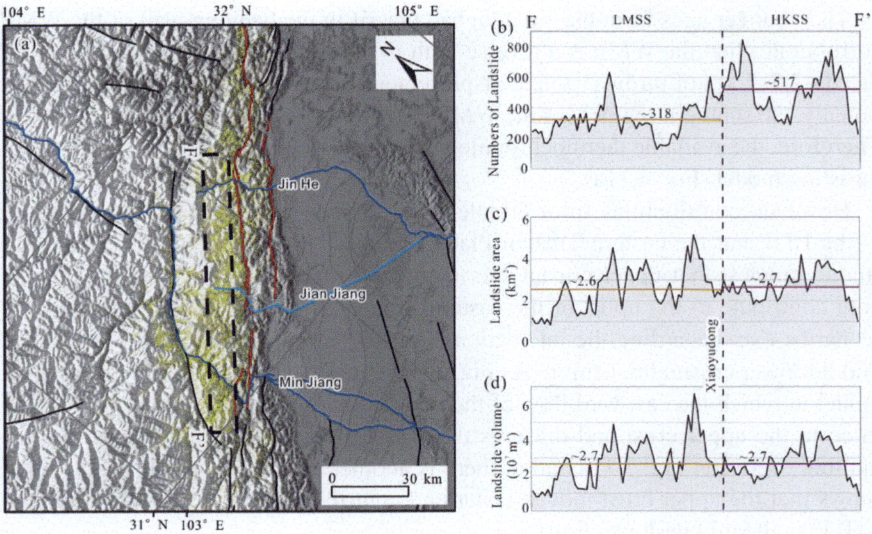

Fig. 3.10 Co-seismic landslides in the central LTB, revised from Tan et al. [43]. **a** Distribution of the landslides (yellow polygons) in the central LTB (data from [56], plotted on the digital elevation model for this region. Also shown are the major rivers (blue), active faults (black), and surface ruptures from the Wenchuan earthquake (red). **b** The number of co-seismic landslides in each sampling window (10 km by 1 km rectangle) along profile F–F′. **c** The area of co-seismic landslides in each sampling window along F–F′. **d** The volume of landslides in each rectangle along F–F′. Orange and pink lines show the mean value of the Longmen and Hongkou sub-segments, respectively

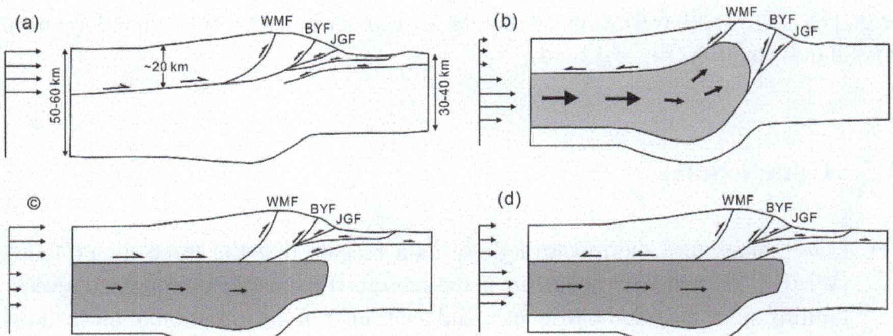

Fig. 3.11 Conceptual structural-tectonic models for the uplift of the eastern Tibetan Plateau margin, modified from Hubbard and Shaw [19], Zhang [65], Tian et al. [46], and Lu et al. [28]. **a** Uplift is entirely produced by upper crustal shortening. **b** Uplift is mainly produced by lower crustal channel extrusion. **c** A modified model between Fig. 3.11a and d. The upper crust moves faster than the ductile lower crust and greatly contributed to mountain building. **d** A modified channel flow model—ductile lower crust flow drags the upper crust eastward to thicken the crust in the LTB through high-angle listric reverse faulting. The arrows on the left side of the Figures represent the horizontal movement of rocks at different depths. Note: Surface topography is exaggerated

The younger ages from the Xuelongbao massif in the hanging wall of the WMF corroborates that the WMF is a reverse fault in the central LTB and has accommodated ~2.5–5 km of throw (vertical displacement) since the middle Miocene. Additionally, the southern segment of the WMF exhibits reverse faulting activity [38, 46]. Therefore, the available thermochronological data largely supports the upper crustal thrusting model (Fig. 3.11a).

However, contributions from middle-lower crustal thickening (or deformation) to the LTB and the eastern Tibetan Plateau cannot be completely refuted. Recent studies suggest that upper crustal shortening and middle-lower crustal thickening both contribute to the uplift of the eastern Tibetan Plateau [41, 65]. Two possible scenarios exist regarding the interaction between the upper crustal shortening and middle-lower crustal thickening. As Fig. 3.11d shows, Zhang et al. [65] proposed a model in which the eastward flow of the ductile lower crust contributes a dragging force to the upper crust and thickens the crust in the LTB through listric reverse faulting. Between Fig. 3.11a and d, there is another model, namely Fig. 3.11c that shows that the upper crust moved faster and contributed more toward building the LTB than the middle-lower crust.

Late-Cenozoic activities of the central LTB major faults indicate that upper crustal thrusting and shortening are one of the major drivers for the mountain building events of the LTB, and that the Pengguan massif is not an uplift block pushed by middle-lower crustal inflation (Fig. 3.11b). Except for the model shown in Fig. 3.11b, the thermochronology results (Fig. 3.6) support all the models presented in Fig. 3.11a–d. An increasing amount of data relating to geophysical observations, such as seismic reflections, receiver functions, Rayleigh wave tomography, and gravity anomalies (e.g., [16, 17, 51, 59, 60]), suggest that the surface exhumation was related to entire crustal deformation (Fig. 3.11c, d).

3.5 Conclusions

(1) Low-temperature thermochronology data reveals that the three major faults (WMF, BYF, and JGF) occurring in the central LTB have accommodated reverse faulting since the middle Miocene, and the timing of the initiation of these faults indicates a mechanism of basinward in-sequence propagation.
(2) The thermochronology data and the seismic reflection profiles demonstrate the along-strike variation of the long-term fault activities. The along-strike variation of fluvial incision intensity in the central LTB is the fundamental driver for the along-strike differences in topography and fault activity.
(3) Late-Cenozoic activity of the major faults in the central LTB indicates that upper crustal thrusting and shortening are major mechanisms for the LTB's uplift. Specifically, uplift of the Pengguan massif is not a result of the middle- to lower crustal flow originating from the Tibetan Plateau. Nevertheless, the contribution of the middle- to lower crust flow to the LTB deformation should be considered.

Acknowledgements This project is supported by the National Natural Science Foundation of China (41302159), the Special Projects for Basic Research Work of the Institute of Geology, China Earthquake Administration (IGCEA1518), and the Ministry of Science and Technology, Taiwan, R.O.C. (MOST 103-2116-M-194-001). We would like to thank Uni-edit (www.uni-edit.net) for editing and proofreading this manuscript.

References

1. Arne D, Worley B, Wilson C, Chen SF, Foster D, Luo ZL, Liu SG, Dirks P (1997) Differential exhumation in response to episodic thrusting along the eastern margin of the Tibetan Plateau. Tectonophysics 280:239–256. https://doi.org/10.1016/S0040-1951(97)00040-1
2. Brandon MT, Roden-Tice M, Garver JI (1998) Late Cenozoic exhumation of the Cascadia accretionary wedge in the Olympic Mountains, northwest Washington State. Geol Soc Am Bull 110:985–1009
3. Burchfiel BC, Chen Z, Liu Y, Royden LH (1995) Tectonics of the Longmen Shan and adjacent regions, central China. Int Geol Rev 37(8):661–735
4. Burchfiel BC, Royden LH, van der Hilst RD, Hager BH, Chen Z, King RW, Li C, Lü J, Yao H, Kirby E (2008) A geological and geophysical context for the Wenchuan earthquake of 12 May 2008, Sichuan, People's Republic of China. GSA Today (Geol Soc Am) 18:4–11
5. Chen Z, Burchfiel BC, Liu Y, King RW, Royden LH, Tang W, Wang E, Zhao J, Zhang X (2000) Global positioning system measurements from eastern Tibet and their implications for India/Eurasia intercontinental deformation. J Geophys Res 105:16,215–16,227. https://doi.org/10.1029/2000jb900092
6. Chen H, Wu Y, Xiao Q (2013) Thermal regime and paleogeothermal gradient evolution of Mesozoic-Cenozoic sedimentary basins in the Tibetan plateau, China. Earth Sci J China Univ Geosci 38(3):541–552
7. Copley A, Avouac JP, Wernicke BP (2011) Evidence for mechanical coupling and strong Indian lower crust beneath southern Tibet. Nature 472(7341):79
8. de Sigoyer J, Vanderhaeghe O, Duchene S, Billerot A (2014) Generation and emplacement of Triassic granitoids within the Songpan Ganze accretionary-orogenic wedge in a context of slab retreat accommodated by tear faulting, Eastern Tibetan plateau, China. J Asian Earth Sci 88:192–216
9. Densmore AL, Ellis MA, Li Y, Zhou RJ, Hancock GS, Richardson N (2007) Active tectonics of the Beichuan and Pengguan faults at the eastern margin of the Tibetan Plateau. Tectonics, 26, TC4005. https://doi.org/10.1029/2006tc001987
10. Donelick RA, O'Sullivan PB, Ketcham RA (2005) Apatite fission-track analysis. Rev Mineral Geochem. https://doi.org/10.2138/rmg.2005.58.3
11. Ehlers TA (2005) Crustal thermal processes and the interpretation of thermochronometer data. Rev Mineral Geochem 58(1):315–350. https://doi.org/10.2138/rmg.2005.58.12
12. England P, Houseman G (1989) Extension during continental convergence, with application to the Tibetan Plateau. J Geophys Res Solid Earth 94(B12):17561–17579
13. England P, Molnar P (1990) Surface uplift, uplift of rocks, and exhumation of rocks. Geology 18:1173–1177
14. Godard V, Pik R, Lavé J, Cattin R, Tibari B, de Sigoyer J, Pubellier M, Zhu J (2009) Late Cenozoic evolution of the central Longmen Shan, eastern Tibet: insight from (U-Th)/He thermochronometry. Tectonics 28. https://doi.org/10.1029/2008TC002407
15. Godard V, Lavé J, Carcaillet J, Cattin R, Bourles D, Zhu J (2010) Spatial distribution of denudation in Eastern Tibet and regressive erosion of plateau margins. Tectonophysics 491:253–274

16. Guo X, Gao R, Keller GR, Xu X, Wang H, Li W (2013) Imaging the crustal structure beneath the eastern Tibetan Plateau and implications for the uplift of the Longmen Shan range. Earth Planet Sci Lett 379:72–80. https://doi.org/10.1016/j.epsl.2013.08.005
17. Guo X, Keller GR, Gao R, Xu X, Wang H, Li W (2014) Irregular western margin of the Yangtze block as a cause of variation in tectonic activity along the Longmen Shan fault zone, eastern Tibet. Int Geol Rev 56(4):473–480
18. Guo X, Gao R, Wang H, Li W, Keller GR, Xu X, Li H, Encarnacion J (2015) Crustal architecture beneath the Tibet-Ordos transition zone, NE Tibet, and the implications for plateau expansion. Geophys Res Lett 42(24):10631–10639. https://doi.org/10.1002/2015GL066668
19. Hubbard J, Shaw JH (2009) Uplift of the Longmen Shan and Tibetan plateau, and the 2008 Wenchuan (M = 7.9) earthquake. Nature 458:194–197
20. Hubbard J, Shaw JH, Klinger Y (2010) Structural setting of the 2008 Mw 7.9 Wenchuan, China, earthquake. Bull Seismol Soc Am 100(5B):2713–2735
21. Jia D, Li Y, Lin A, Wang M, Chen W, Wu X, Ren Z, Zhao Y, Luo L (2010) Structural model of 2008 Mw 7.9 Wenchuan earthquake in the rejuvenated Longmen Shan thrust belt, China. Tectonophysics 491(1–4):174–184. https://doi.org/10.1016/j.tecto.2009.08.040
22. Kirby E, Reiners PW, Krol MA, Whipple KX, Hodges KV, Farley KA, Tang W, Chen Z (2002) Late Cenozoic evolution of the eastern margin of the Tibetan Plateau: inferences from 40Ar = 39Ar and (U-Th)/He thermochronology. Tectonics 21. https://doi.org/10.1029/2000TC001246
23. Li K, Xu XW, Tan XB, Chen GH, Xu C, Kang WJ (2015) Late quaternary deformation of the Longquan anticline in the Longmenshan thrust belt, eastern Tibet, and its tectonic implication. J Asian Earth Sci 112:1–10
24. Li Z, Liu S, Chen H, Deng B, Hou M, Wu W, Cao J (2012) Spatial variation in Meso-Cenozoic exhumation history of the Longmen Shan thrust belt (eastern Tibetan Plateau) and the adjacent western Sichuan basin: constraints from fission track thermochronology. J Asian Earth Sci 47:185–203
25. Liang S, Gan W, Shen C, Xiao G, Liu J, Chen W, Ding X, Zhou D (2013) Three-dimensional velocity field of present-day crustal motion of the Tibetan Plateau derived from GPS measurements. J Geophys Res: Solid Earth 118(10):5722–5732
26. Liu Q, van der H. Robert Y, Li HJ, Yao JH, Chen B, Guo SH, Qi J, Wang H, Li SC (2014) Eastward expansion of the Tibetan Plateau by crustal flow and partitioning across faults. Nat Geosci https://doi.org/10.1038/NGEO2130
27. Liu-Zeng J, Zhang Z, Wen L, Tapponnier P, Sun J, Xing X, Hu G, Xu Q, Zeng L, Ding L, Ji C, Hudnut KW, van der Woerd J (2009) Co-seismic ruptures of the 12 May 2008, Ms 8.0 Wenchuan earthquake, Sichuan: east-west crustal shortening on oblique, parallel thrusts along the eastern edge of Tibet. Earth Planet Sci Lett 286(3–4):355–370
28. Lu R, He D, John S, Wu JE, Liu B, Chen Y (2014) Structural model of the central Longmen Shan thrusts using seismic reflection profiles: Implications for the sediments and deformations since the Mesozoic. Tectonophysics 630:43–53
29. Molnar P, Boos WR, Battisti DS (2010) Orographic controls on climate and paleoclimate of Asia: thermal and mechanical roles for the Tibetan Plateau. Ann Rev Earth Planet Sci 38
30. Morley CK (1988) Out-of-sequence thrusts. Tectonics 7(3):539–561
31. Richardson NJ, Densmore AL, Seward D, Fowler A, Wipf M, Ellis MA, Yong L, Zhang Y (2008) Extraordinary denudation in the Sichuan Basin: insights from low-temperature thermochronology adjacent to the eastern margin of the Tibetan Plateau. J Geophys Res Solid Earth 113(B4)
32. Robert A, Zhu J, Vergne J, Cattin R, Chan L-S, Wittlinger G, Herquel G, de Sigoyer J, Pubellier M, Zhu LD (2010) Crustal structures in the area of the 2008 Sichuan earthquake from seismologic and gravimetric data. Tectonophysics 491:205–210
33. Royden LH, Burchfiel BC, King RW, Wang E, Chen Z, Shen F, Liu Y (1997) Surface deformation and lower crustal flow in eastern Tibet. Science 276(5313):788–790
34. Shen ZK, Sun J, Zhang P, Wan Y, Wang M, Burgmann R, Zeng Y, Gan W, Liao H, Wang Q (2009) Slip maxima at fault junctions and rupturing of barriers during the 2008 Wenchuan earthquake. Nat Geosci. https://doi.org/10.1038/NGEO636

35. Shi F, He HL, Wei ZY (2012) Coseismic horizontal shortening associated with the 2008 Wenchuan Earthquake along the Bashahe segment from high resolution satellite images. J Asian Earth Sci. https://doi.org/10.1016/j.jseaes.2012.01.001
36. Sun C, Jia D, Yin H, Chen Z, Li Z, Shen L, Wei D, Li Y, Yan B, Wang M, Fang S, Cui J (2016) Sandbox modeling of evolving thrust wedges with different preexisting topographic relief: Implications for the Longmen Shan thrust belt, eastern Tibet. J Geophys Res Solid Earth. https://doi.org/10.1002/2016jb013013
37. Sun M, Yin A, Yan D, Ren H, Mu H, Zhu L, Qiu L (2018) Role of pre-existing structures in controlling the Cenozoic tectonic evolution of the eastern Tibetan plateau: new insights from analogue experiments. Earth Planet Sci Lett 491:207–215. https://doi.org/10.1016/j.epsl.2018.03.005
38. Tan XB, Lee YH, Chen WY, Cook KL, Xu XW (2014) Exhumation history and faulting activity of the southern segment of the Longmen Shan, eastern Tibet. J Asian Earth Sci 81:91–104
39. Tan XB, Xu XW, Lee YH, Yuan RM, Yu GH, Xu C (2015) Differential Late-Cenozoic vertical motions of the Beichuan-Yingxiu Fault and the Jiangyou-Guanxian Fault in the central Longmenshan range and their tectonic implications. Chin J Geophys 58(1):143–152 (in Chinese with English abstract)
40. Tan XB, Xu XW, Lee YH, Lu RQ, Liu Y, Xu C, Li K, Yu GH, Kang WJ (2017) Late Cenozoic thrusting of major faults along the central segment of Longmen Shan, eastern Tibet: evidence from low-temperature thermochronology. Tectonophysics 712:145–155
41. Tan X, Xu X, Lu R (2017) Comment on "Sandbox modeling of evolving thrust wedges with different preexisting topographic relief: implications for the Longmen Shan thrust belt, eastern Tibet. J Geophys Res Solid Earth. https://doi.org/10.1002/2016jb013505
42. Tan XB, Lee YH, Xu XW, Cook KL (2017) Cenozoic exhumation of the Danba Antiform, Eastern Tibet: evidence from low-temperature thermochronology. Lithosphere. https://doi.org/10.1130/L613.1
43. Tan X, Yue H, Liu Y, Xu X, Shi F, Xu C, Ren Z, Shyu JBH, Lu R, Hao H (2018) Topographic loads modified by fluvial incision impact activity in the Longmenshan thrust belt, eastern margin of the Tibetan plateau. Tectonics. https://doi.org/10.1029/2017TC004864
44. Tapponnier P, Molnar P (1976) Slip-line field theory and large-scale continental tectonics. Nature 264(5584):319
45. Tapponnier P, Zhiqin X, Roger F, Meyer B, Arnaud N, Wittlinger G, Jingsui Y (2001) Oblique stepwise rise and growth of the Tibet Plateau. Science 294(5547):1671–1677
46. Tian Y, Kohn BP, Gleadow AJW, Hu S (2013) Constructing the Longmen Shan eastern Tibetan Plateau margin: insights from low-temperature thermochronology. Tectonics 32(3):576–592
47. Wang E, Kirby E, Furlong KP, van Soest M, Xu G, Shi X, Kamp PJJ, Hodges KV (2012) Two-phase growth of high topography in eastern Tibet during the Cenozoic. Nat Geos https://doi.org/10.1038/NGEO1538
48. Wang M, Jia D, Shaw JH, Hubbard J, Plesch A, Li Y, Liu B (2014) The 2013 Lushan earthquake: implications for seismic hazards posed by the range front blind thrust in the Sichuan Basin. China, Geology 42(10):915–918
49. Wang Y, Ma J, Li C (2008) The migration characteristics of strong earthquakes on the North-South seismic belt and its relation with the south Asia seismic belt. Earthq Res China 22(3):237–250
50. Wang X, Zhang G, Fang H, Luo W, Zhang W, Zhong Q, Cai X, Luo H (2014) Crust and upper mantle resistivity structure at middle section of Longmenshan, eastern Tibetan plateau. Tectonophysics 619–620:143–148
51. Wei X, Jiang M, Liang X, Chen L, Ai Y (2017) Limited southward underthrusting of the Asian lithosphere and material extrusion beneath the northeastern margin of Tibet, inferred from teleseismic Rayleigh wave tomography. J Geophys Res Solid Earth 122:7172–7189. https://doi.org/10.1002/2016JB013832
52. Whipple KX, Meade BJ (2004) Controls on the strength of coupling among climate, erosion, and deformation in two-sided, frictional orogenic wedges at steady state. J Geophys Res 109:F01011

53. Willett SD, Brandon MT (2002) On steady states in mountain belts. Geology 30:175–178
54. Wilson CJ, Fowler AP (2011) Denudational response to surface uplift in east Tibet: Evidence from apatite fission-track thermochronology. Bulletin 123(9–10):1966–1987
55. Wilson CJL, Harrowfield MJ, Reid AJ (2006) Brittle modification of Triassic architecture in eastern Tibet: implications for the construction of the Cenozoic plateau. J Southeast Asian Earth Sci 27:341–357
56. Xu C, Xu X, Yao X, Dai F (2014) Three (nearly) complete inventories of landslides triggered by the May 12, 2008 Wenchuan Mw 7.9 earthquake of China and their spatial distribution statistical analysis. Landslides 11(3):441–461
57. Xu M, Zhu CQ, Tian YT, Rao S, Hu SB (2011) Borehole temperature logging and characteristics of subsurface temperature in the Sichuan Basin. Chin J Geophys 54(4):1052–1060 (In Chinese)
58. Xu X, Wen X, Yu G, Chen G, Klinger Y, Hubbard J, Shaw JH (2009) Coseismic reverse- and oblique-slip surface faulting generated by the 2008 Mw 7.9 Wenchuan earthquake. China Geol 27:515–518
59. Xu X, Gao R, Guo X, Li W, Li H, Wang H, Huang X, Lu Z (2017) Outlining tectonic inheritance and construction of the Min Shan region, eastern Tibet, using crustal geometry. Sci Rep 7(1):13798. https://doi.org/10.1038/s41598-017-14354-4
60. Xu X, Hong Y, Zhou Q, Liu J, Yuan L, Wang J (2018) Century-scale high-resolution black carbon records in sediment cores from the South Yellow Sea, China. J Oceanol Limnol 36(1):114–127
61. Yan DP, Zhou MF, Li SB, Wei GQ (2011) Structural and geochronological constraints on the Mesozoic-Cenozoic tectonic evolution of the Longmen Shan thrust belt, eastern Tibetan Plateau. Tectonics 30. https://doi.org/10.1029/2011TC002867
62. Yin A (2010) A special issue on the great 12 May 2008 Wenchuan earthquake (Mw7. 9): observations and unanswered questions. Tectonophysics 1(491):9
63. Yin A, Harrison TM (2000) Geologic evolution of the Himalayan-Tibetan orogen. Ann Rev Earth Planet Sci 28:211–280
64. Yu G, Xu X, Klinger Y, et al (2010) Fault-scarp features and cascading-rupture model for the Wenchuan earthquake (Mw 7.9), Eastern Tibetan Plateau, China. Bull Seismol Soc America 100(5B). https://doi.org/10.1785/0120090255
65. Zhang PZ (2013) A review on active tectonics and deep crustal processes of the Western Sichuan region, eastern margin of the Tibetan Plateau. Tectonophysics 584:7–22
66. Zhang Z, Yuan X, Chen Y, Tian X, Kind R, Li X, Teng J (2010) Seismic signature of the collision between the east Tibetan escape flow and the Sichuan Basin. Earth Planet Sci Lett 292(3–4):254–264
67. Zhao G, Unsworth MJ, Zhan Y, Wang L, Chen X, Jones AG, Tang J, Xiao Q, Wang J, Cai J, Li T, Wang Y, Zhang J (2012) Crustal structure and rheology of the Longmenshan and Wenchuan Mw 7.9 earthquake epicentral area from magnetotelluric data. Geology 40(12):1139–1142

Chapter 4
A Review of the 1999 Chi-Chi, Taiwan, Earthquake from Modeling, Drilling, and Monitoring with the Taiwan Chelungpu-Fault Drilling Project

Kuo-Fong Ma

Abstract Many high-quality strong motion stations were built prior to the occurrence of the destructive 1999 Chi-Chi earthquake, providing the most comprehensive study of the mechanism of such a damaging event. The general consistent feature in spatial slip distribution of the fault as a large slip of ~12 m at the northern portion of the fault from fault models and geological observation suggest the importance in understanding the physics of faulting of this large slip. The Taiwan Chelungpu-fault Drilling Project (TCDP) aided in the understanding of earthquake energy partitions by revealing the very fine grain (~nm) of the fault gouge with a millimeter-scale slip thickness for a single event. The dynamic parameters obtained from the kinematic slip inversion suggest a heterogeneous shear stress distribution and a complex stress-time history. The study incorporated the examination of the surface energy from identifying the fault gouge from the major slip zone, and the determination of the fracture energy from the dynamic parameters derived from the strong motion data. This allowed for the direct estimation of the energy partition of a single earthquake from the geological and seismological observations. The low-frictional coefficient obtained from temperature measurements made after drilling encouraged similar drilling projects to be undertaken after large earthquakes (e.g. 2008 Wenchuan and 2011 Tohoku earthquakes) to obtain frictional heating measurements. With the success of the TCDP drilling in the slip zone associated with the 1999 Chi-Chi earthquake, in situ borehole seismometers, referred to as the TCDP borehole seismic array (TCDPBHS), were installed to monitor the behavior of the fault zone after a large slip. This paper shows the dynamic parameters obtained from the kinematic slip inversion and, thus, gives a brief review of the current understanding of earthquake

K.-F. Ma (✉)
Earthquake-Disaster & Risk Evaluation and Management (E-DREaM) Center, National Central University, Zhongli, Taoyuan, Taiwan
e-mail: fong@earth.sinica.edu.tw

Department of Earth Sciences, National Central University, Zhongli, Taoyuan, Taiwan

Present Address:
Institute of Earth Sciences, Academia Sinica, Taipei, Taiwan

C.-H. Lo et al. (eds.), *Earthquake Geology and Tectonophysics around Eastern Tibet and Taiwan*, Atmosphere, Earth, Ocean & Space,
https://doi.org/10.1007/978-981-15-6210-5_4

kinematics and dynamics. Seismic waveform modeling, drilling of the Chelungpu-fault, and the in situ monitoring of the fault zone after a large slip of ~12 m are all discussed. As a large number of related scientific papers have been published on this subject, some of these papers may not be properly addressed or referred to in this review article. These papers may be identified in the references of related articles.

4.1 Introduction

The study of earthquake dynamics is essential to understand the underlying physics of earthquakes, as well as their initiation, propagation, and healing. The constitutive relation of the fault slip is the key to dynamically describe seismic sources. This constitutive relation was assumed to be in the form of a slip-weakening model [2, 3, 6, 13], which is completely characterized by the initial stress, yield stress, dynamic frictional stress, and slip-weakening distance. More complicated constitutive laws might be required to understand the re-strengthening of faults that occur during ruptures. In addition, several studies that modeled large earthquakes, namely the 1992 Landers, 1994 Northridge, and the 1995 Kobe earthquakes, showed the significant heterogeneous stress fields and spatially varying friction in relation to the earthquake rupture.

The exceptionally large slip and slip velocity observed in the 1999 Chi-Chi earthquake have been of note when studying the physics of earthquake ruptures [26]. Detailed studies of the waveform inversion of this earthquake have utilized the excellent near-field, far-field, and GPS data shown in Fig. 4.1 (e.g. [16, 26]), and comprehensive modeling has been performed using three-dimensional geometries and velocity structures [19, 22, 20]. These kinematic studies show the spatio-temporal heterogeneous distribution of the slip over the fault plane. It was shown that the majority of the slip occurred in the northern portion of the fault with a slip of ~12–14 m obtained from most of the slip models. Ma et al. [29] suggested that the large slip and slip velocity within the fault is a result of fault lubrication of the rupture dynamics during faulting.

A deep hole scientific drilling program (Dakeng well, 2000 m), named the Taiwan Chelungpu-fault Drilling Project (TCDP), was carried out in 2004 [27] after two shallow holes were successfully drilled to the north (Fengyuan well, 450 m) and south (Nantou well, 220 m) of the Chelungpu-fault [41], as shown in Fig. 4.2. This program aimed to aid in the understanding of the physical conditions of the 1999 Chi-Chi earthquake, and it provides a rare opportunity to observe the fault rupture at different depths. This provides insight into the: (1) continuous core profile of a large displacement rupture at a depth where the elastic strain energy was stored and released; (2) sample material in the high-speed slip zone; and (3) determine the physical conditions (stress, pore pressure, temperature) within the fault zone. The TCDP drilling program drilled two holes (Hole-A and Hole-B) separated by 20 m with a sidetrack, as shown in Fig. 4.3. Primary slip zones associated with the 1999 Chi-Chi earthquake were identified at depths of 1111 and 1137 m, respectively, for

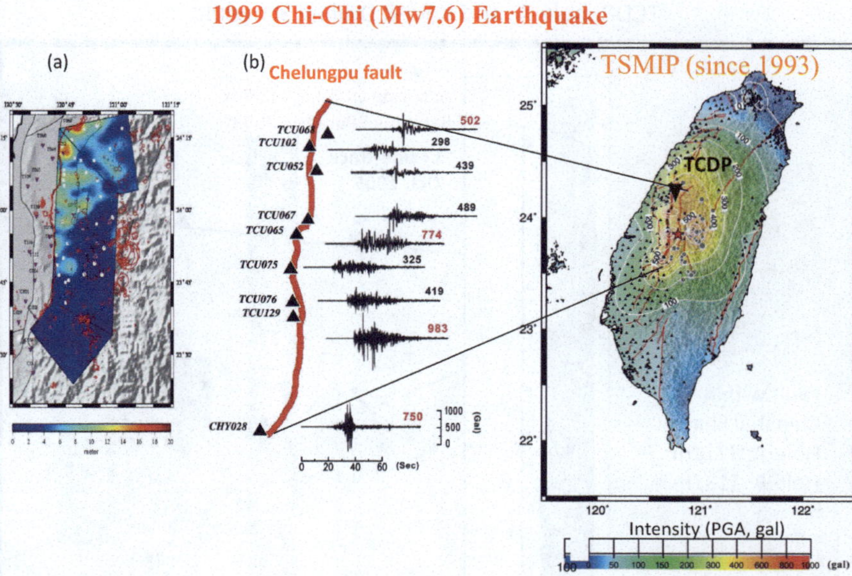

Fig. 4.1 The 1999 Chi-Chi earthquake. **a** The ruptured Chelungpu-fault and the spatial slip distribution from [16]. **b** The distribution of the Taiwan strong-motion implementation program (TSMIP, since 1993), the near-fault strong-motion stations with recorded accelerograms, and the location of the Taiwan Chelungpu-fault drilling program (TCDP) drill site

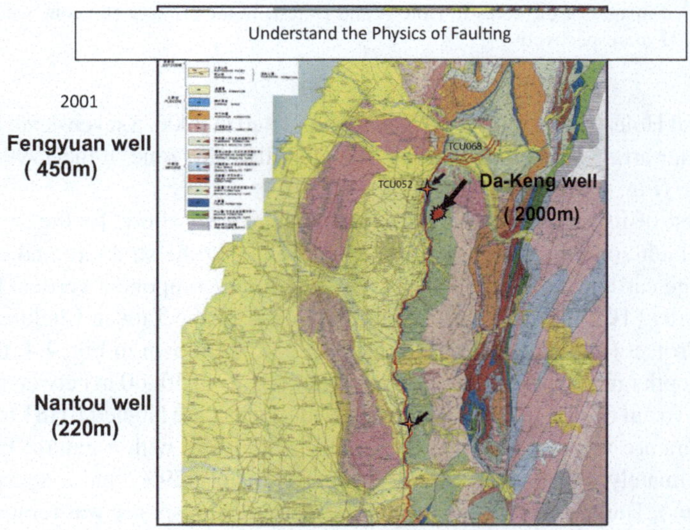

Fig. 4.2 A geological map and the locations of two shallow holes (indicated by the red crosses) at the Fengyuan (450 m) and Naotou (220 m) sites to the north and south of the Chelungpu-fault, respectively. The location of the TCDP 2000 m hole at Dakeng is also shown

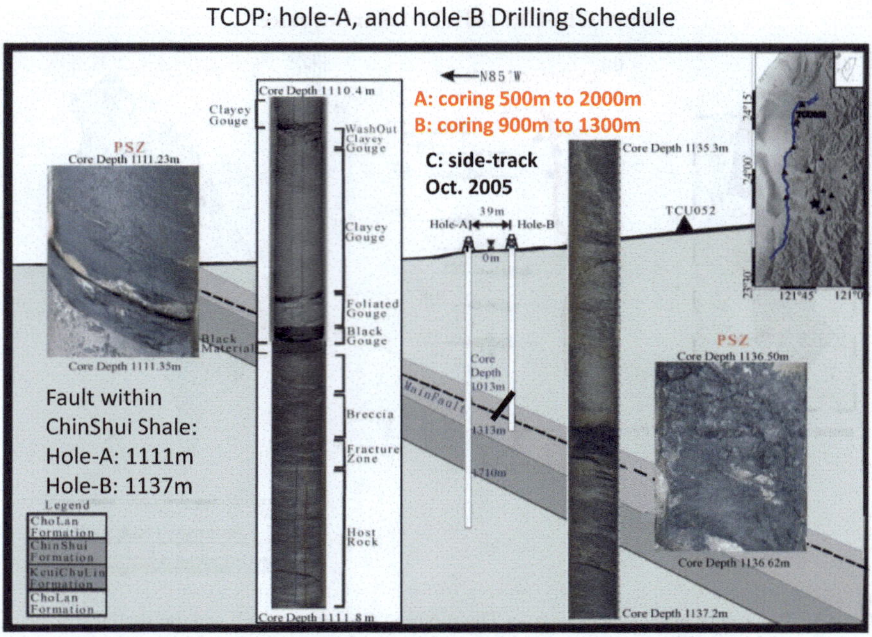

Fig. 4.3 The layout of the Taiwan Chelungpu-fault drilling program (TCDP) with two deep holes, Hole-A and Hole-B, and a side-track Hole-C. Images show the 1 m cores close to the identified primary slip zone (indicated by the dark color in the imaged core) for the 1999 Chi-Chi earthquake. The magnified images of the cores in Hole-A and Hole-B in the primary slip zone were taken at 1111 and 1137 m, respectively

Hole-A and Hole-B. After the drilling was completed in 2006, a seven-level borehole seismometer array was installed across the identified slip zone of the 1999 Chi-Chi earthquake (Fig. 4.4).

This first-of-its-kind in situ borehole seismometer array allows for the close observation of fault zone dynamics to understand the fault zone structure and evolution after a large earthquake. The TCDP seven-level three-component vertical borehole seismic array (TCDPBHS) was installed in Hole-A of the Taiwan Chelungpu-fault Drilling Project (TCDP) in July 2006 [22, 21, 25]. As shown in Fig. 4.4, this array covers a depth range from 946 to 1274 m with intervals of 50–60 m between the seismometers (from the top to the bottom, the seismometers are labeled BHS1 to BHS7). The sensors are velocity-type short-period seismometers with a natural frequency of approximately 4.5 Hz, a damping of approximately 29%, and a sensitivity of 1.6 V/(cm/s). The recording gain is 100. The instrument response was removed from each record, and corrections for the Galperin angle and orientation of the three-component sensors were performed systematically before waveform analysis [22, 21]. The sampling rate was set at 1000 samples/s to detect micro-earthquakes with high frequency signals up until the end of 2007.

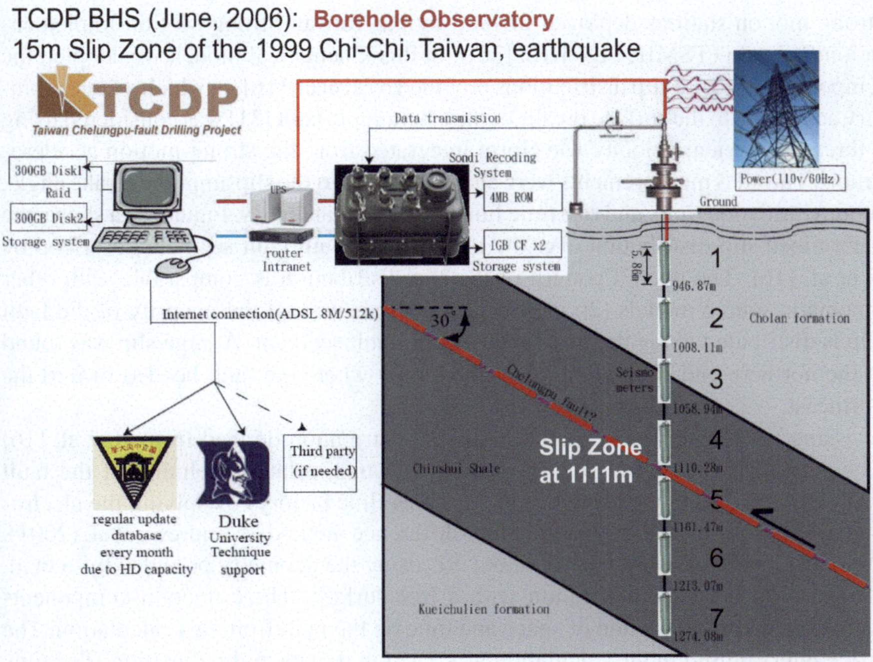

Fig. 4.4 The layout of the three-component, seven-level vertical borehole seismic array situated across the identified primary slip zone with the first borehole seismometer, BHS1, at a depth of around 950 m. Distances of about 50 m separated each seismometer until a depth of around 1300 m, and these were named BHS2-BHS7. BHS4 was placed near the identified primary slip zone at 1111 m

This paper gives a brief review of the understanding of earthquake kinematics and dynamics from seismic waveform modeling, drilling of the Chelungpu-fault, and the in situ monitoring of the fault zone after a large slip of ~12 m. As a large number of scientific papers were published in relation to this subject, it is possible that some of these important papers were not properly addressed or referred to in this review article. However, these related publications might be noted in the references of related articles.

4.2 Estimation of the Dynamic Parameters from the Kinematic Modeling of the 1999 Chi-Chi Earthquake

Kinematic modeling of the 1999 Chi-Chi, Taiwan earthquake was carried out to provide a temporal-spatial slip distribution using data obtained from the high-quality strong motion stations deployed in 1993 by the Taiwan Strong Motion Implementation Program (TSMIP, Fig. 4.1b, [36]). In this kinematical model, we adopted the temporal and spatial slip distributions provided by Ji et al. [16], in which a fault geometry according to the surface breaks of the Chelungpu-fault [31] was considered using a three-component velocity waveform integrated from the strong-motion accelerograms. The GPS measurements were also used to invert the slip amplitude, rake angle, rupture initiation time, and rise time functions simultaneously. Figure 4.2a shows the determined slip distribution over the main north–south fault segment described by Ji et al. [16]. The overall pattern of the slip distribution is comparable with other kinematic source models [26, 48]. This model shows that the majority of the fault slip is distributed along the main south–north fault segment. A large slip was found at the northern end of the fault, near the region where the fault bended toward the northeast.

Using the temporal and spatial fault slip distribution derived from Ji et al. [16] as constraints (Fig. 4.5a), we determined the stress-related evolution of the fault that occurred during the rupture (i.e. the stress-time history) by solving the electrodynamic equations in the dynamic finite-difference method of Andrews et al. (2001). We considered a dipping fault of about 30° using the geometry provided by Ji et al. [16] in a heterogeneous medium with a free surface. Three traction components were obtained as a function of space and time on the fault from this calculation. The space interval used in the calculation was 2 km in the down-dip direction. The time step was 0.0109 s, which was chosen to satisfy the stability in the finite-difference calculation. The velocity structure is a 1D central Taiwan model [28], which was used in the kinematic modeling of the earthquake by Ji et al. [16].

Figure 4.5b, c shows the changes in the final shear and normal tractions. The change in the final shear traction is rather heterogeneous as the stress increased and decreased by as much as 40 MPa. This region surrounds a large slip and shows a large increase in the stress after rupture. The change in the final normal traction was about 4 MPa, which is quite negligible compared to the change in shear traction. Thus, for the stress-time history investigation only the shear stress was considered in this study. The stress-time history curve shows the changes in the stress of the subfaults during fault rupture. This curve was obtained using the dynamic finite-difference calculations and is shown in Fig. 4.6 for each of the subfaults at a down-dip depth of less than 16 km for the southern portion of the fault. Likewise, these curves were calculated for the central segment (Fig. 4.7) and northern portion of the fault (Fig. 4.8) for the region exhibiting a large slip. The regions with lesser amounts of slip in comparison to their surroundings showed significant stress increases after the fault ruptured, as shown in Fig. 4.6. The stress-time history beyond the depth of the hypocenter showed significantly different dynamic and static stress drops. For the subfault at the hypocenter, a dynamic stress decrease of about 25 MPa and a static stress decrease of about 15 MPa were observed. Some subfaults showed that the stress initially decreased where the rupture front passed through the subfault, but underwent a process of re-strengthening during the slip, as shown in Fig. 4.7. The shear stress in the region that experienced a large slip did not reach a steady-state, suggesting the possible involvement of a dynamic behavior such as lubrication

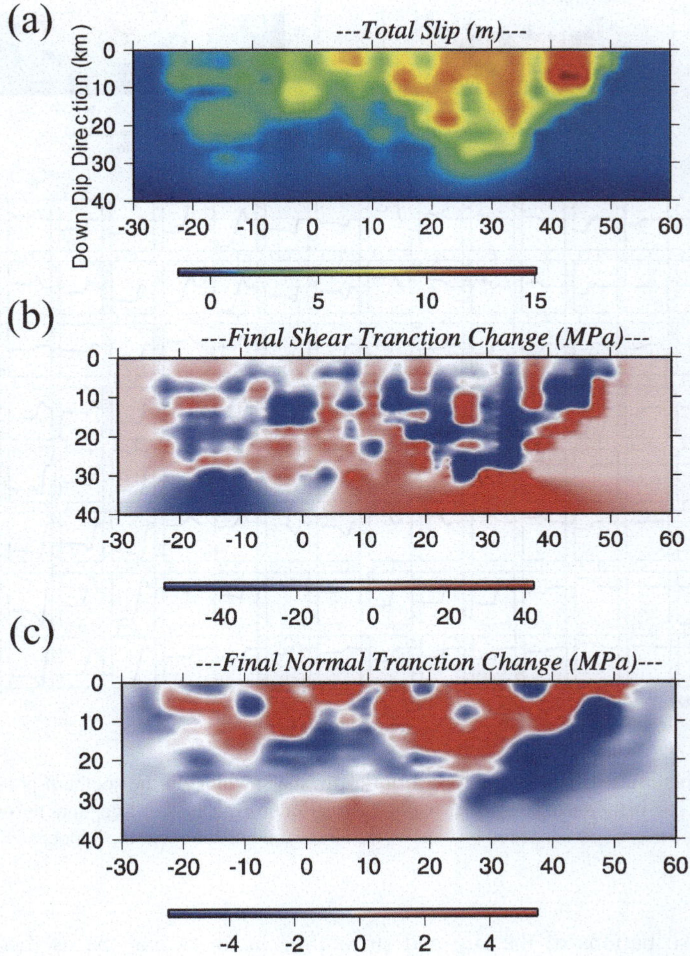

Fig. 4.5 **a** The spatial slip distribution of the main north–south segment provided by Ji et al. [16]. The inverted final **b** shear traction and **c** normal traction distribution over the fault plane obtained from the dynamic parameter model

during faulting [29]. The above analysis of the dynamic parameters suggests that the dynamic behavior of the propagation of the fault rupture significantly affects the final amount of slip. It remains difficult, however, to understand the detailed dynamic rupture behavior in relation to the final amount of slip due to the lack of information on the initial stress and strength of the fault. Peyrat et al. [35] showed that seismic data cannot distinguish between barriers, and models of the asperities must assume the existence of either a variable initial stress or a variable yield stress (strength) over the fault. The heterogeneous distribution of the strength values, shown in Figs. 4.6, 4.7, and 4.8, might also indicate the heterogeneity of the initial stress.

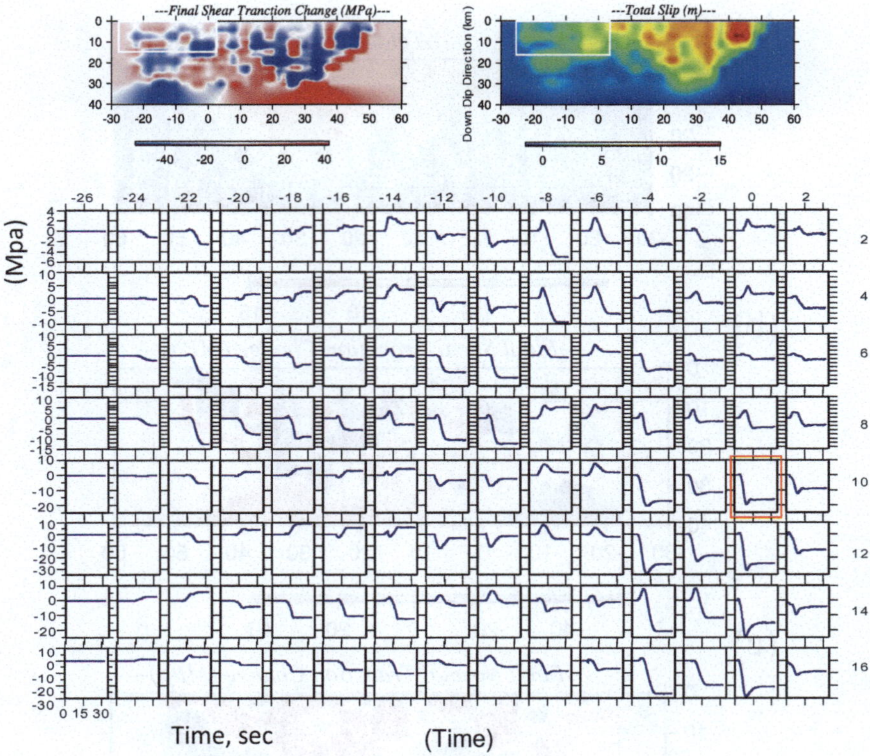

Fig. 4.6 The inverted shear stress-time histories for the subfaults in the southern portion of the fault (white box in the shear stress and spatial slip distribution maps) over a depth range of 2–16 km and evaluated every 2 km. The red box shows the stress-time history near the hypocenter at a depth of 10 km

The distributions of the slip and stress-time histories may be used to infer a constitutive relation at each grid point in space. Ide and Takeo [14] examined the constitutive relations of a fault slip for the 1995 Kobe earthquake. They found significant variations in the slip-weakening rates between the shallow and deep portions of the fault. Although the resolution of the present study might be limited by the numerical calculations that are used, we also showed the possible depth dependence of the slip-weakening rate by plotting the stress and slip at each time step up to 33 s after the rupture initiation for the regions near the hypocenter and middle (Fig. 4.9), as well as the northern portion of the fault with the large-slip region (Fig. 4.10). These plots show the variation of the apparent slip-weakening curves with the depth. For the majority of the subfaults, the stress decreases as the slip increases (slip-weakening), as has been observed in frictional experiments of rocks in laboratory. Overall, it was shown that the slip-weakening rates at shallow depths (<6 km in the dip direction) are significantly smaller than those at greater depths (10–14 km in the dip direction) regardless of the location along the fault. At the hypocenter, the

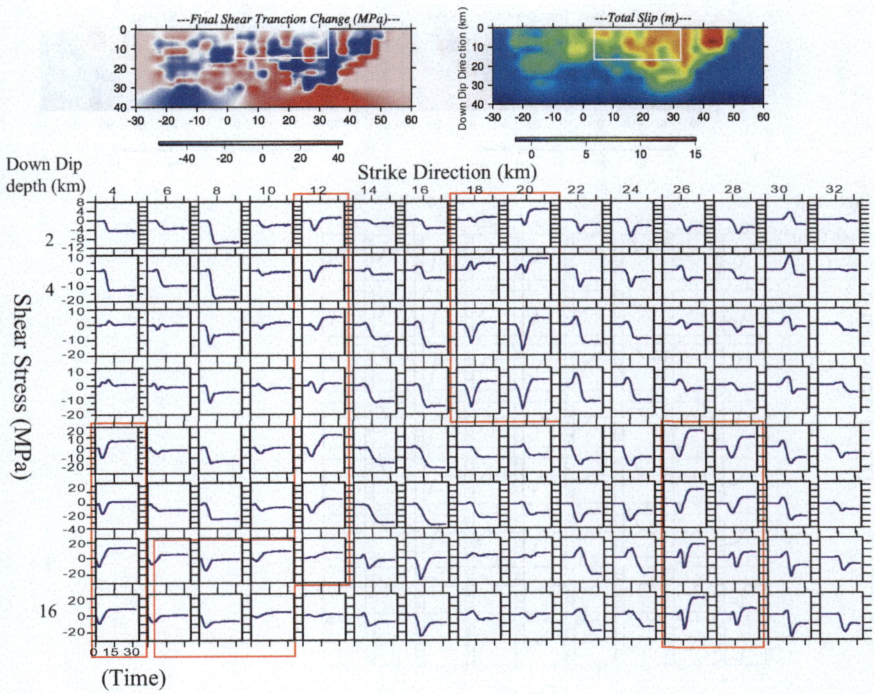

Fig. 4.7 Using the same conditions as in Fig. 4.6, but concerning the central section of the fault. The subfaults are marked by red boxes, indicating the regions with increased final stress levels after the slip via stress strengthening

stress decreases by about 2 MPa at the shallow depths during a slippage of about 4–6 m, while the stress decreases by 12–25 MPa during a slippage of about 6–8 m. For the middle region of the fault, wherein the slip extends to a maximum depth of 30 km in the dip direction, the slip-weakening rate at shallow depths is also much smaller than those at greater depths. Except for the subfault at a depth of 22 km, the slip-weakening rates do not differ significantly at greater depths. However, the slip-weakening rate tends to increase as the depth increases. The value of Dc, calculated where the amount of slip with the stress decreases to the least, decreased with increasing depth for depths greater than 6 km. This Dc value, however, is an apparent slip-weakening distance (Dc^a) as it might be influenced by the discrete data in the numerical analysis, and thus differs from the Dc defined in experiments involving rock mechanics. For the slip-weakening curves in the region with large slips (40–46 km north of the hypocenter, Fig. 4.10), the slip-weakening rates at shallow depths are again smaller than those at greater depths. Comparatively, the slip-weakening rates of the subfaults near the northern end of the fault vary less with the depth. In general, it is still shown that the slip-weakening rate increases at greater depths. With the exception of the subfaults near the northern end of the fault, the plots of these slip-weakening curves show that the slip-weakening rates at shallow depths are

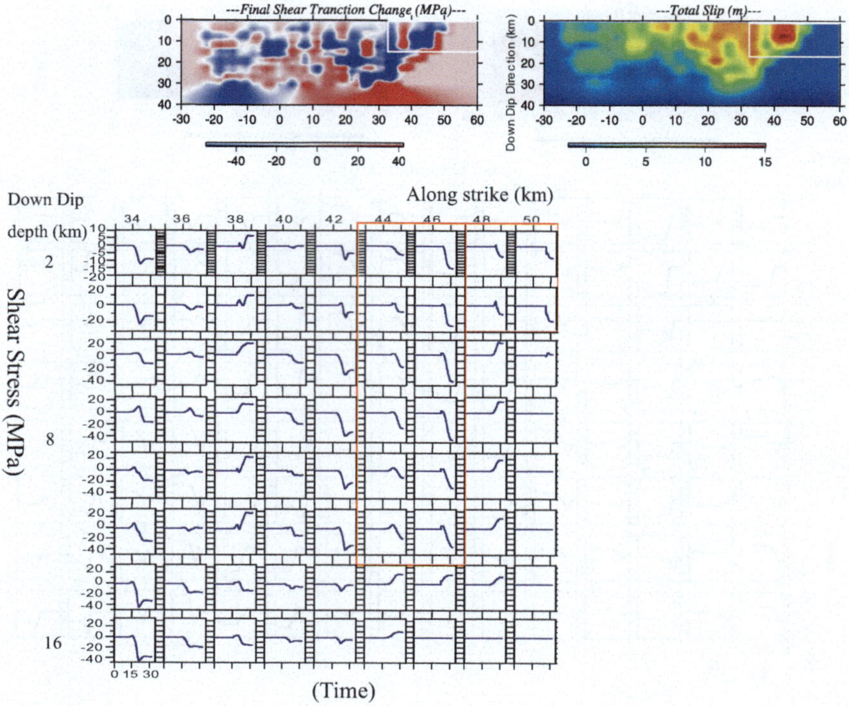

Fig. 4.8 Using the same conditions as Fig. 4.6, but concerning the northern section of the fault, which experienced a large slip of ~12 m. The subfaults are marked with red boxes and show that the shear stress did not reach a steady state after slipping

Fig. 4.9 The inverted stress–slip curves (apparent slip-weakening curves) over a depth range of 2–16 km and evaluated every 2 km, as indicated by different colors. The curves at the hypocenter (0 km) and at 30 km north of the hypocenter are shown

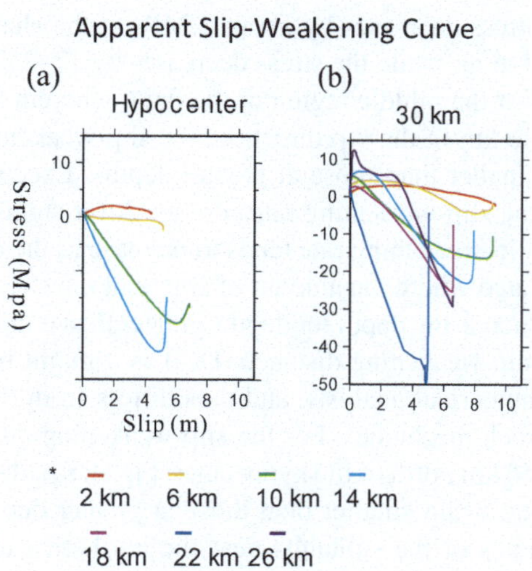

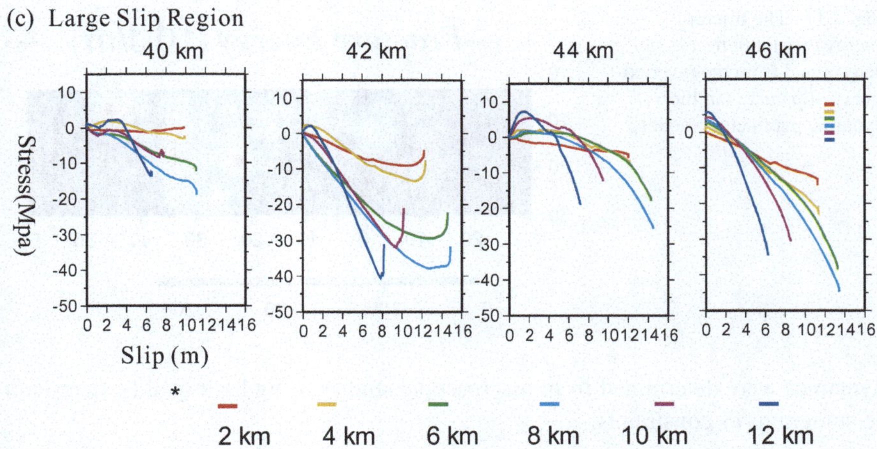

Fig. 4.10 The inverted stress–slip curves (apparent slip-weakening curves) for regions that experienced large slipping over a depth range of 2–16 km and evaluated every 2 km. These regions were located at 40, 42, 44, and 46 km north of the hypocenter

significantly smaller than those at greater depths, while the slip-weakening curves at shallow depths demonstrated similar slip rates. This indicates that the obtained apparent slip-weakening distance, Dc^a, is larger at shallower depths than at greater depths. Overall, the slip-weakening rate was found to increase with the depth, as indicated by Ide and Takeo [14] for the 1995 Kobe earthquake. This previous study suggests that the differences in the slip-weakening rates are too large to be explained by the existence of low rigidity at shallow depths, and similar features were found between the 1999 Chi-Chi earthquake and the 1995 Kobe earthquake.

To determine the distribution of the apparent slip-weakening distance, Dc^a, over the fault from the slip-weakening curve considering the limited frequency resolution of the kinematic model, this Dc^a could be considered the upper bound of the actual values. While the determination of Dc might be limited by numerical calculation, the determination of the fracture energy from this dynamic model is achievable, as suggested by Guatteri and Spudich [8]. The fracture energy is defined as the amount of energy needed to create a unit-area of fault via shear fracture. The distribution of the fracture energy is estimated from the apparent slip-weakening curve and the excess of strength, excluding the subfaults with a negative stress drop, and is shown in Fig. 4.11. Estimates of the shear-fracture energy have been calculated for many earthquakes (e.g. [1, 4, 12, 13, 32, 34]), yielding a range from 10^2 to 10^8 J/m^2. The fracture energy we estimated near the hypocenter for this event was about 10^8 J/m^2. The regions corresponding to large strength excesses have larger fracture energies of up to 2–3 × 10^8 J/m^2. The determination of the facture energy and the slip-weakening curve developed from seismological data provide first-hand information that aids in the understanding of the earthquake energy partition during faulting. These earthquake

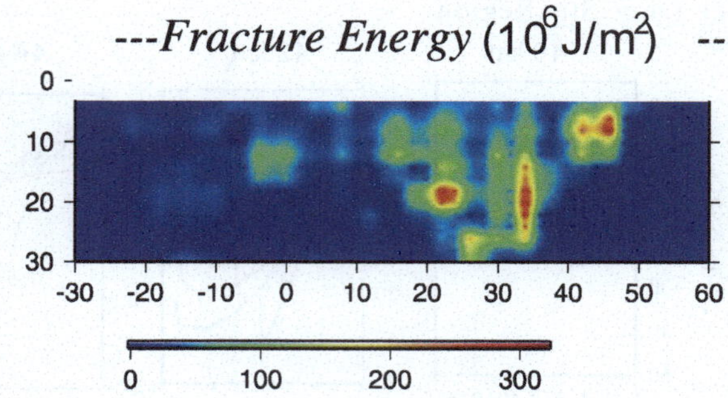

Fig. 4.11 The fracture energy distribution determined from the inverted stress–slip curves of the dynamic parameters model

dynamics were determined from macroscopic studies using high-quality, near-fault seismograms as constraints.

4.3 Taiwan Chelungpu-Fault Drilling Project (TCDP)

The unique scientific objective of the TCDP is to understand an earthquake energy budget [27]. It incorporates a macroscopic analysis of the seismic dynamic parameter model stated above, a microscopic study of the fault gouge surface fracture energy [27], and an analysis of the residual heat following drilling [17, 41–39]. More studies have been carried out to examine the nano-spherule grains retrieved from the fault core [10, 18]. The large slip and slip velocity in the northern portion of the fault have been considered to have resulted from thermal pressurization faulting dynamics [9] that suggested for the existence of a fluid [15], and a discussion on the thermal parameters across the fault was undertaken by Tanaka et al. [38, 39]. The fault zone permeability [7, 33] was suggested as a key parameter in this process. For the 1999 Chi-Chi earthquake, the stress before and after the earthquake has also been of interest. The earthquake focal mechanisms with the thrust mechanism of the mainshock induced mostly strike-slip aftershocks, as calculated using Coulomb stress changes [30] and studies of the aftershocks (e.g. [46]). The FMI and DSI [11, 23, 47] geophysical logs were used to show the rotation of the stress axes. The horizontal shear stress before the earthquake is at its maximum value, dropping to a minimum after the earthquake, as indicated and modeled by a leak-off test that demonstrated this decrease in stress [47].

This scientific drilling in a large slip zone after an earthquake was successfully carried out to determine the fault gouge associated with the 1999 Chi-Chi earthquake. The actual thickness of the zone that slips during the rupture of a large earthquake was, thus, identified and found to be on the scale of a few millimeters (Fig. 4.12). This is a key seismological parameter used to understand energy dissipation, rupture processes, and seismic efficiency. The cores retrieved from the Taiwan Chelungpu-fault Drilling Project, the main slip zone associated with the Chi-Chi earthquake, and the surface fracture energy estimated from the grain sizes in the gouge zone of the

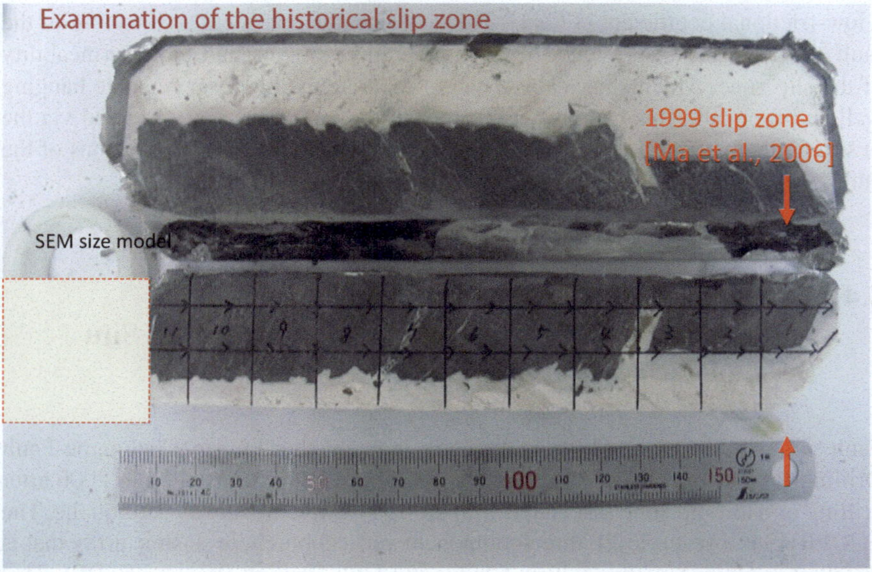

Fig. 4.12 The polished primary slip zone [27] and sample preparation by Kuniyo Kawabata (personal communication) for further historical slip zone studies. The major slip zone was identified at the bottom of the primary slip zone as the least crossing crack with a thickness of about 5 mm for the 1999 Chi-Chi earthquake

fault sample were directly compared to the seismic fracture energy. The seismic fracture energy was determined from the above dynamic parameters and the temporal–spatial slip distribution of the kinematic inversion obtained from the dense near-field strong-motion data. From this comparison, it was shown that 6% of the earthquake breakdown work was contributed by the gouge surface energy [27].

The retrieved fault gouge had small grain sizes of 50–100 mm in the primary slip zone and images showed grain sizes of less than 50 nm with rounded shapes. This suggests that these small grains might be the result of precipitation rather than fracturing (Fig. 4.12). A thermal-pressurization process was proposed to be the mechanism of the large slip in the Chi-Chi earthquake based on the evidence found for the existence of fluid during this earthquake [15]. This indicates that a rapid temperature rise of hundreds of degrees in this thin slip zone (on the order of a millimeter–centimeter thickness) may induce a slip of the order of 10 m. The composition of the fault gouge material suggests that it was full of clay minerals [37]. The on-site estimation of the hydrological parameters and rock experiments involving the fault gouge material suggested that they possess a low permeability of the order of 10^{-20} m^2, regardless of the high permeability of the damage zone in the hanging wall (10^{-17} m^2) and the high permeability and porosity of the footwall in the Kuichulin formation (obtained from the shallow hole studies, [40]). The in situ temperature measurement suggests a very low temperature residual from the identified slip zone, and suggests a low-frictional coefficient [17, 41]. The geological/geophysical architecture of the fault core suggests the existence of very fine grains (~50 nm) and a low permeability of the slip zone, which behaves as a cap to separate the highly permeable hanging wall and footwall. This geophysical feature of the fault zone was observed via the in situ fault zone borehole seismometers that provided close-up observations of the fault zone evolution that occurs after a large earthquake.

4.4 Determination of the Fault Zone Behavior and Evolution After a Large Earthquake from In Situ Seismological Observations

Fault zone in situ borehole seismometers, namely the Taiwan Chelungpu-Fault Drilling Project Borehole Seismometers (TCDPBHS), were installed in 2006 after drilling into the slip zone associated with the 1999 Chi-Chi, Taiwan earthquake. The TCDPBHS is a seven-level, three-component vertical borehole seismic array that is installed in Hole-A of the Taiwan Chelungpu-fault Drilling Project (TCDP). This array covers a depth range of 946–1274 m with seismometers at intervals of 50–60 m, and it crosses the major slip zone of this earthquake at a depth of 1111 m. The TCDPBHS observed no micro-seismicity in the fault zone (Fig. 4.13a) as a result from the total stress drop [22], but observed occurrences of isotropic events below the primary slip zone [25], as shown in the TCDP profile with the lithology structure and velocity log in Fig. 4.13b. These isotropic events were explained as events with an explosive/implosive mechanism that is driven by the fluid that forms during the total decrease in stress and exists in a region capped by a low-permeability primary slip zone. This suggests that the fluid in the fault zone plays a significant role in the evolution of the fault zone. Through almost a decade of observations made with the TCDPBHS, significant and intriguing features had been discovered. The study of the relationship between the seismic moment, M_o, and the source duration, tw, of the micro-earthquake clusters recorded by the TCDPBHS showed high waveform similarity at a constant duration for event magnitudes ranging from 0.3 to 2.0 within the clusters. These events mainly occurred in the décollement at depths of about 10–15 km. To determine how M_o scales with tw for these event clusters, path effects were examined using a path-averaged Q value as determined from the deconvolution of the vertical seismic array of the event cluster. The results indicated a nearly constant tw for events within each cluster, regardless of M_o. This constant duration may arise because all events in a cluster are hosted on the same, isolated seismogenic patch, or because the events are driven by external factors of constant duration, such as fluid injections into the décollement. It may also be related to the size of the earthquake nucleation. Although there was a constant tw, we also observed some differences in the frequency content for various M_o values, suggesting that a dynamic rupture pattern rather than a simple circular fault might take place during slipping, as proposed by Lin and Lapusta [24] via dynamic modeling.

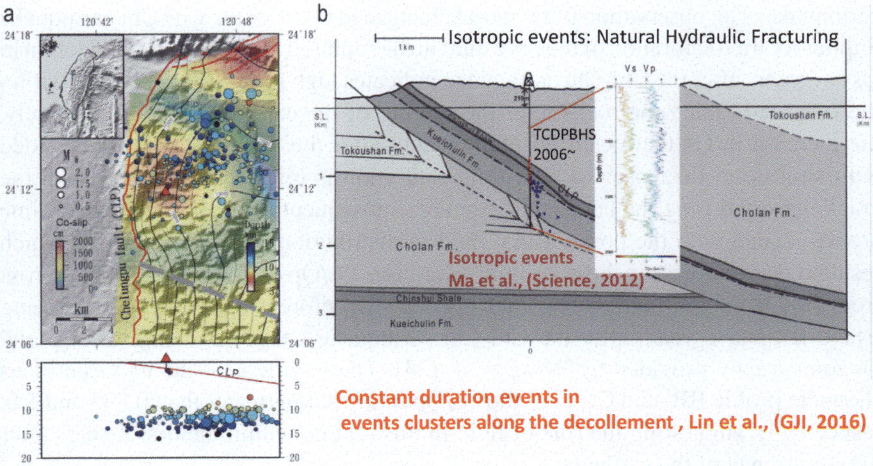

Fig. 4.13 A summary of the **a** TCDP site (red triangle), showing no aftershocks along the Chelungpu-fault (CLP) with seismicity from depths of about 10–15 km along the décollement [22], and **b** TCDP profile with the lithology structure, borehole seismometers, and velocity log of the TCDP. The hydraulic parameters from the core sample analysis of the retrieved fault cores revealed low permeability of the primary slip zone and high permeability for the damaged hanging wall. High porosity and high permeability were noted for the Kuechulin formation of the footwall. The total stress drop of the Chi-Chi earthquake was noted at this location from FMI analysis [47], and the detection of the isotropic events [25] at depths of about 1300–1500 m within Kuechulin formation were found

The attenuation factor, Q, at the fault zone is an important parameter for understanding the physical properties of this zone. Using the seven-level TCDP borehole seismometer array placed in the hanging wall and footwall of the identified slip zone, the Q-structure of the recently ruptured fault zone was determined. After analyzing several micro-events with incident angles of less than 40°, Wang et al. [42] obtained a low Qs of ~21–22 and Qp of ~27–35 for the damage zone in the hanging wall. These Qp and Qs values are quite consistent with the observations of the San Andreas fault made at a corresponding depth. A low Qs value for the recent Chelungpu-fault zone suggests that this fault zone has been highly fractured. The Qs values within the Chelungpu-fault that are similar to those within the San Andreas fault suggest that the Q structure within the fault zone is sedimentary-rock-independent. However, the possible existence of fluids, fractures, and cracks dominates the attenuation feature in the fault zone.

The role of fluid and the evolution of the fault zone could be investigated indirectly through analysis of the attenuation factors (Q^{-1}), as stated above. In addition to the fault zone Q structure analysis, a decrease in Qs within the fault rupture zone that occurred two years after the 1999 Chi-Chi earthquake was revealed by Qs tomography images and an analysis of the single-path Qs near the Chelungpu-fault [45]. Along with a 1% Vs reduction, revealed by the analysis of repeating earthquakes [5], this suggested a possible reduction in both Vs and Qs within the fault zone after a large earthquake. The observation of temporal changes in Qs after the Chi-Chi earthquake implies a varied saturation of the pore fluid in the ruptured fault zone. The reduction in Qs two years after the Chi-Chi earthquake indicates high pore-fluid saturation within the fractured fault zone rocks after the rupture of the earthquake. More precisely, the increase in Qs that occurred immediately after the aftershocks that coincided with short-term decreases in pore-fluid saturation is related to an increase in the crack density during the co-seismic rupture. Subsequent decreases in Qs over time are associated with the post seismic fluid re-distribution and crack healing, which resulted in an increase in the pore-fluid saturation. Qs/Qp tomography profiles across from the Chelungpu-fault are shown in Fig. 4.14 for before (stage I), 2 years following (stage II), and 2 years after the Chi-Chi earthquake (stage III) using the Qp and Qs tomography provided by Wang et al. [43]. The regime near the hypocenter, as shown in profile BB′ and CC′, has the Qs/Qp ratio values greater than 0.833 and 1.0, respectively, suggesting the role of fluid in this regime within the first 2 years after the occurrence of the earthquake.

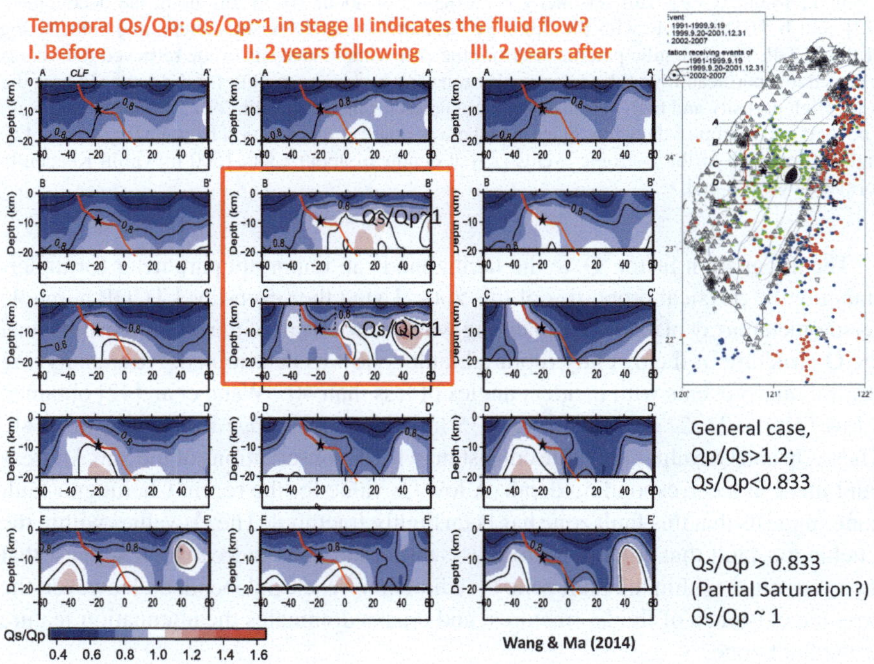

Fig. 4.14 The Qs/Qp profiles of AA′–EE′ across the Chelungpu-fault from Qs and Qp tomography [43] shown for before (stage I), 2 years following (stage II), and 2 years after (stage III) the 1999 Chi-Chi earthquake, with seismicity indicated by different colors. The red line indicates the Chelungpu-fault at a depth delineated by the seismic reflection experiment [44]. The dashed box in the profile CC′ shows the region with the most significant change in Qs from a previous study [42], which used data from stage III. A contour with Qs/Qp > 0.833 was also found, indicating the possible involvement of fluid as shown by the red boxes along the profile BB′ and CC′ at stage II, 2 years after the earthquake

4.5 Conclusions

The dense population of strong-motion stations that were implemented prior to the occurrence of the 1999 Chi-Chi, Taiwan earthquake provided the opportunity for extensive study of this earthquake, despite the massive destruction it caused. This data improved our understanding of destructive earthquakes through this high-quality, near-fault seismological data. The largest in-land slip (~12 m) occurred near the surface and provided a unique opportunity to understand the faulting physics during a large slip via drilling. The thickness of the slip zone was of the order of millimeters and allowed for significant evidence of a single event to be obtained through this project [25]. The distinct low-frictional coefficient from in situ temperature measurement was also an important discovery to resolve the paradox on the strong or weak of the fault in the crust [17, 41]. This low-frictional coefficient discovered during TCDP brought the doubt on the thermal dissipation, and a workshop to address this topic for a rapid response drilling of ICDP (International Continental Drilling Program) was proposed in 2009 (https://websites.pmc.ucsc.edu/~rapid/). The Taiwan Chelungpu-fault drilling was an international collaborated program with the ICDP (International Continental Drilling Program). It also involved the experiences of Japanese colleagues in regards to the Nojima fault, a fault that is associated with the 1995 Kobe earthquake, as well as their findings via drilling and concerns of possible Nankai megathrust earthquakes [38, 39]. The TCDP provided the direct observation of the physical and chemical behaviors of faulting through on-site experiments and analysis of the retrieved core, and these findings have been published worldwide. The experience of the TCDP encouraged other drilling projects, including the 2008 Wenchuan fault scientific drilling program (WFSD) and the 2011 J-FAST drilling program for the 2011 Tohoku earthquake. These modern scientific drilling programs advance our understanding of earthquake faulting dynamics. Through years of observation provided by the TCDPBHS, we continue to reveal the fault zone dynamics via close-up seismological observations. Through the in situ borehole seismometers placed in the fault zone of a large earthquake after its occurrence, close-up observations of fault zone behavior were attained and its temporal evolution was elucidated.

References

1. Aki K (1979) Characterization of barriers on an earthquake fault. J Geophys Res Solid Earth 84(B11):6140–6148
2. Andrews DJ (1976) Rupture propagation with finite stress in antiplane strain. J Geophys Res 81(20):3575–3582
3. Andrews DJ (1976) Rupture velocity of plane strain shear cracks. J Geophys Res 81(32):5679–5687
4. Beroza GC, Spudich P (1988) Linearized inversion for fault rupture behavior: application to the 1984 Morgan Hill, California, earthquake. J Geophys Res Solid Earth 93(B6):6275–6296

5. Chen KH, Furumura T, Rubinstein J, Rau RJ (2011) Observations of changes in waveform character induced by the 1999 Mw7. 6 Chi-Chi earthquake. Geophys Res Lett 38(23)
6. Day SM (1982) Three-dimensional simulation of spontaneous rupture: the effect of nonuniform prestress. Bull Seismol Soc Am 72(6A):1881–1902
7. Doan ML, Brodsky EE, Kano Y, Ma KF (2006) In situ measurement of the hydraulic diffusivity of the active Chelungpu Fault, Taiwan. Geophys Res Lett 33(16)
8. Guatteri M, Spudich P (2000) What can strong-motion data tell us about slip-weakening fault-friction laws? Bull Seismol Soc Am 90(1):98–116
9. Hirono T, Sakaguchi M, Otsuki K, Sone H, Fujimoto K, Mishima T, Yeh EC (2008) Characterization of slip zone associated with the 1999 Taiwan Chi-Chi earthquake: X-ray CT image analyses and microstructural observations of the Taiwan Chelungpu fault. Tectonophysics 449(1–4):63–84
10. Hirono T, Yeh EC, Lin W, Sone H, Mishima T, Soh W, Kinoshita M (2007) Nondestructive continuous physical property measurements of core samples recovered from hole B, Taiwan Chelungpu-fault drilling project. J Geophys Res Solid Earth 112(B7)
11. Hung JH, Ma KF, Wang CY, Ito H, Lin W, Yeh EC (2009) Subsurface structure, physical properties, fault-zone characteristics and stress state in scientific drill holes of Taiwan Chelungpu fault drilling project. Tectonophysics 466(3–4):307–321
12. Husseini MI, Jovanovich DB, Randall MJ, Freund LB (1975) The fracture energy of earthquakes. Geophys J Int 43(2):367–385
13. Ida Y (1974) Slow-moving deformation pulses along tectonic faults. Phys Earth Planet Inter 9(4):328–337
14. Ide S, Takeo M (1997) Determination of constitutive relations of fault slip based on seismic wave analysis. J Geophys Res Solid Earth 102(B12):27379–27391
15. Ishikawa T, Tanimizu M, Nagaishi K, Matsuoka J, Tadai O, Sakaguchi M, Kikuta H (2008) Coseismic fluid–rock interactions at high temperatures in the Chelungpu fault. Nat Geosci 1(10):679
16. Ji C, Helmberger DV, Wald DJ, Ma KF (2003) Slip history and dynamic implications of the 1999 Chi-Chi, Taiwan, earthquake. J Geophys Res Solid Earth 108(B9)
17. Kano Y, Mori J, Fujio R, Ito H, Yanagidani T, Nakao S, Ma KF (2006) Heat signature on the Chelungpu fault associated with the 1999 Chi-Chi, Taiwan earthquake. Geophys Res Lett 33(14)
18. Kuo LW, Song SR, Yeh EC, Chen HF (2009) Clay mineral anomalies in the fault zone of the Chelungpu Fault, Taiwan, and their implications. Geophys Res Lett 36(18)
19. Lee SJ, Ma KF (2000) Rupture process of the 1999 Chi-Chi, Taiwan, earthquake from the inversion of teleseismic data. Terrest Atmos Ocean Sci 11(3):591–608
20. Lee SJ, Chen HW, Ma KF (2006). Three-dimensional dense strong motion waveform inversion for the rupture process of the 1999 Chi-Chi, Taiwan, earthquake. J Geophys Res Solid Earth, 111(B11)
21. Lin YY, Ma KF, Kanamori H, Song TRA, Lapusta N, Tsai VC (2016) Evidence for non-self-similarity of microearthquakes recorded at a Taiwan borehole seismometer array. Geophys J Int 206(2):757–773
22. Lin YY, Ma KF, Oye V (2012) Observation and scaling of microearthquakes from the Taiwan Chelungpu-fault borehole seismometers. Geophys J Int 190(1):665–676
23. Lin W, Yeh EC, Ito H, Hung JH, Hirono T, Soh W, Song SR (2007) Current stress state and principal stress rotations in the vicinity of the Chelungpu fault induced by the 1999 Chi-Chi, Taiwan, earthquake. Geophys Res Lett 34(16)
24. Lin YY, Lapusta N (2018) Microseismicity simulated on asperity-like fault patches: on scaling of seismic moment with duration and seismological estimates of stress drops. Geophys Res Lett. https://doi.org/10.1029/2018GL078650
25. Ma KF, Lin YY, Lee SJ, Mori J, Brodsky EE (2012) Isotropic events observed with a borehole array in the Chelungpu fault zone, Taiwan. Science 337(6093):459–463
26. Ma KF, Mori J, Lee SJ, Yu SB (2001) Spatial and temporal distribution of slip for the 1999 Chi-Chi, Taiwan, earthquake. Bull Seismol Soc Am 91(5):1069–1087

27. Ma KF, Tanaka H, Song SR, Wang CY, Hung JH, Tsai YB, Sone H (2006) Slip zone and energetics of a large earthquake from the Taiwan Chelungpu-fault drilling project. Nature 444(7118):473
28. Ma KF, Wang JH, Zhao D (1996) Three-dimensional seismic velocity structure of the crust and uppermost mantle beneath Taiwan. J Phys Earth 44(2):85–105
29. Ma KF, Brodsky EE, Mori J, Ji C, Song TRA, Kanamori H (2003) Evidence for fault lubrication during the 1999 Chi-Chi, Taiwan, earthquake (Mw7. 6). Geophys Res Lett 30(5)
30. Ma KF, Chan CH, Stein RS (2005) Response of seismicity to Coulomb stress triggers and shadows of the 1999 Mw = 7.6 Chi-Chi, Taiwan, earthquake. J Geophys Res Solid Earth, 110(B5)
31. Ma KF, Song TRA, Lee SJ, Wu HI (2000) Spatial slip distribution of the September 20, 1999, Chi-Chi, Taiwan, earthquake (Mw7. 6)—inverted from teleseismic data. Geophys Res Lett 27(20):3417–3420
32. Mikumo T, Olsen KB, Fukuyama E, Yagi Y (2003) Stress-breakdown time and slip-weakening distance inferred from slip-velocity functions on earthquake faults. Bull Seismol Soc Am 93(1):264–282
33. Murakami H, Hashimoto T, Oshiman N, Yamaguchi S, Honkura Y, Sumitomo N (2001) Electrokinetic phenomena associated with a water injection experiment at the Nojima fault on Awaji Island, Japan. Island Arc 10(3–4):244–251
34. Olsen KB, Madariaga R, Archuleta RJ (1997) Three-dimensional dynamic simulation of the 1992 Landers earthquake. Science 278(5339):834–838
35. Peyrat S, Olsen K, Madariaga R (2001) Dynamic modeling of the 1992 Landers earthquake. J Geophys Res Solid Earth 106(B11):26467–26482
36. Shin TC, Kuo KW, Lee WHK, Teng TL, Tsai YB (2000) A preliminary report on the 1999 Chi-Chi (Taiwan) earthquake. Seismol Res Lett 71(1):24–30; Tanaka H, Chen WM, Kawabata K, Urata N (2007) Thermal properties across the Chelungpu fault zone and evaluations of positive thermal anomaly on the slip zones: are these residuals of heat from faulting? Geophys Res Lett 34(1)
37. Song SR, Kuo LW, Yeh EC, Wang CY, Hung JH, Ma KF (2007) Characteristics of the lithology, fault-related rocks and fault zone structures in TCDP Hole-A. Terr Atmos Ocean Sci 18(2):243–269
38. Tanaka H, Chen WM, Kawabata K, Urata N (2007) Thermal properties across the Chelungpu fault zone and evaluations of positive thermal anomaly on the slip zones: are these residuals of heat from faulting? Geophys Res Lett. https://doi.org/10.1029/2006GL028153
39. Tanaka H, Chester FM, Mori JJ, Wang C-Y (2007) Drilling into the fault zones. Tectonophyiscs 443:123–290
40. Tanaka H, Wang CY, Chen WM, Sakaguchi A, Ujiie K, Ito H, Ando M (2002) Initial science report of shallow drilling penetrating into the Chelungpu fault zone, Taiwan. Terr Atmos Ocean Sci 13:227–251
41. Tanaka H, Chen WM, Wang CY, Ma KF, Urata N, Mori J, Ando M (2006) Frictional heat from faulting of the 1999 Chi-Chi, Taiwan earthquake. Geophys Res Lett 33(16)
42. Wang YJ, Lin YY, Lee MC, Ma KF (2012) Fault zone Q values derived from Taiwan Chelungpu Fault borehole seismometers (TCDPBHS). Tectonophysics 578:76–86
43. Wang YJ, Ma KF, Mouthereau F, Eberhart-Phillips D (2010) Three-dimensional Qp and Qs-tomography beneath Taiwan orogenic belt: implications for tectonic and thermal structure. Geophys J Int 180:891–910
44. Wang CY, Li CL, Yen HY (2002) Mapping the northern portion of the Chelungpu fault, Taiwan by shallow reflection seismics. Geophys Res Lett 29(16):1790. https://doi.org/10.1029/2001GL014496
45. Wang YJ, Ma KF (2015) Investigation of the temporal change in attenuation within the ruptured fault zone of the 1999 Mw7. 3 Chi-Chi, Taiwan earthquake. Pure Appl Geophys 172(5):1291–1304
46. Wu YM, Zhao L, Chang CH, Hsu YJ (2008) Focal-mechanism determination in Taiwan by genetic algorithm. Bull Seismol Soc Am 98(2):651–661

47. Wu H-Y, Ma K-F, Zoback M, Boness N, Ito H, Hung J-H, Hickman S (2007) Stress orientations of Taiwan Chelungpu-Fault Drilling Project (TCDP) hole-A as observed from geophysical logs. Geophys Res Lett 34(1):L01303
48. Zeng Y, Chen CH (2001) Fault rupture process of the 20 September 1999 Chi-Chi, Taiwan, earthquake. Bull Seismol Soc Am 91(5):1088–1098

Chapter 5
Seismotectonics of the 2013 Lushan M_W 6.7 Earthquake—A Hidden Earthquake

Ren-Qi Lu, Xi-Wei Xu, Deng-Fa He, John Suppe, Xi-Bin Tan, Ying-Qiang Li, and Zhen-Nan Wang

Abstract Almost five years after the devastating May 12, 2008, M_W 7.9 Wenchuan earthquake, southern Longmen Shan was struck by the M_W 6.7 Lushan earthquake on April 20, 2013. After the Lushan earthquake, it was debated whether the fault was a high-angle outcrop fault or a low-angle blind fault. The geometry, kinematics, and seismotectonic model of this earthquake has also been debated widely. In our study, several high-resolution artificial seismic reflection profiles are combined with near-surface geological data and well-drilling data to determine the sedimentary and structural deformation of the Lushan area. Our study integrates the focal mechanism solution, aftershock relocation, and GPS and leveling data to reveal the coseismic fault and seismotectonics of the 2013 Lushan earthquake. Three-dimensional imaging of the aftershocks is used to identify two planar faults that form a Y-shape (f1 and f2). The main active fault f1 dips to the northwest at approximately 45°–50° at depths of 7–19 km. Seismic interpretations suggest that it did not break through the overlying Mesozoic and Cenozoic rocks, and also that it is a blind fault. Geodetic measurements suggest that the coseismic deformation is consistent with the geometry and kinematics of shear fault-bend folding. The deep seismic reflection profile indicates that the syndepositional nature of the fault is a pre-existing normal fault older than the Triassic that underwent positive inversion tectonics during the Late Cenozoic. The 2013 Lushan earthquake, which was triggered by blind faults, is deemed to be a hidden earthquake. Blind and reactivated faults increase the potential risk and uncertainty related to earthquakes in the eastern margin of the Tibetan Plateau.

R.-Q. Lu (✉) · X.-B. Tan · Y.-Q. Li · Z.-N. Wang
State Key Laboratory of Earthquake Dynamics, Institute of Geology, China Earthquake Administration, Beijing, China
e-mail: Lurq127@sina.com

X.-W. Xu
Institute of Crustal Dynamics, China Earthquake Administration, Beijing, China

D.-F. He
College of Energy Resources, China University of Geosciences, Beijing, China

J. Suppe
Department of Earth and Atmospheric Sciences, University of Houston, Houston, TX, USA

C.-H. Lo et al. (eds.), *Earthquake Geology and Tectonophysics around Eastern Tibet and Taiwan*, Atmosphere, Earth, Ocean & Space,
https://doi.org/10.1007/978-981-15-6210-5_5

Keywords Blind fault · Hidden earthquake · Seismotectonics · Shear fault-bend folding · Inversion tectonics · Lushan earthquake

5.1 Introduction

On April 20, 2013, an unexpected M_w 6.7 earthquake occurred in Lushan County at the southern side of Longmen Shan (LMS). The hypocenter of this earthquake was located approximately 80 km southwest of the 2008 Wenchuan earthquake (Fig. 5.1). Both earthquakes resulted in a large number of casualties and losses [36, 37]. Some studies have suggested that the seismogenic fault was a high-angle listric-reverse fault that is probably linked to the Shuangshi-Dachuan fault (F4) [4]; however, a field

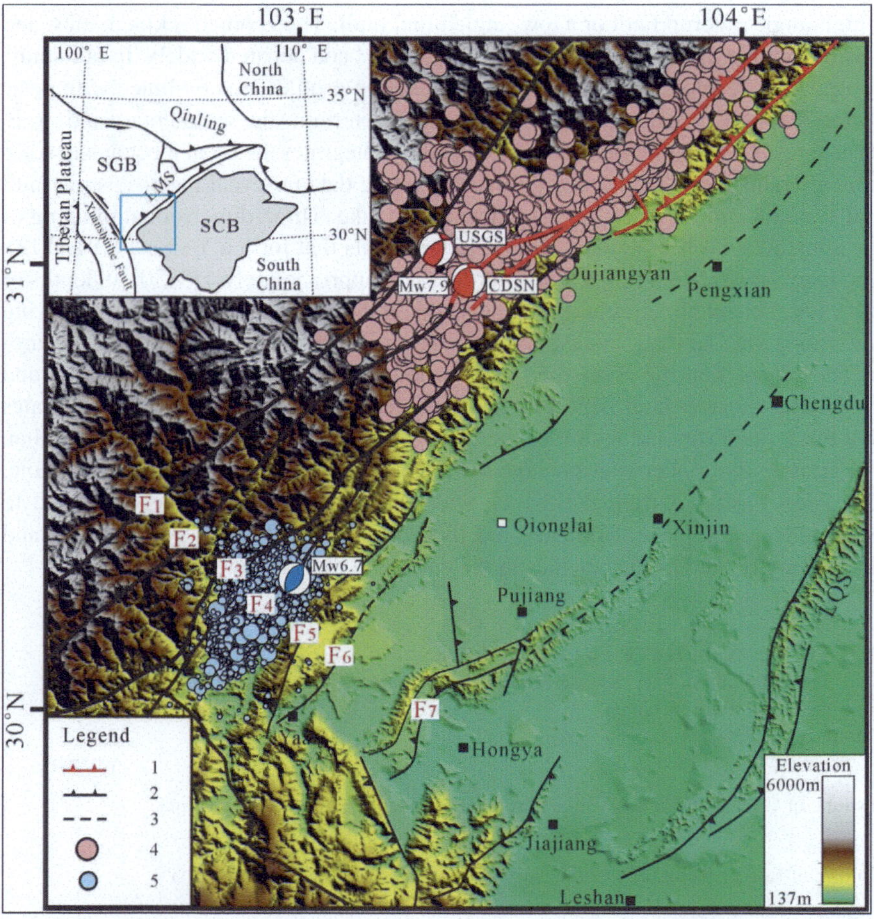

Fig. 5.1 The major fault system and earthquakes in the southern LMS thrust belt. The black solid lines represent the major faults and the dashed lines represent the blind or concealed faults

investigation did not find coseismic surface ruptures [16] and did not support the idea that F4 is the seismogenic fault. Seismic and geological interpretations suggest that the seismogenic fault is a gently dipping frontal blind thrust at an approximate angle of 30° [17, 33]; however, the relocated aftershocks and focal mechanism solution indicate that the seismogenic fault dips at ~7° to the northwest [9, 41]. In addition, other geological studies have suggested that the Lushan earthquake was caused by thrusting and detachment folding in the frontal propagation belt of the LMS [21] and that the seismogenic fault was the Dayi fault (F6) in the frontal mountain [18]. Moreover, geodetic measurements, including GPS and leveling data obtained during the earthquake, suggest a maximum coseismic displacement in the Lushan syncline area [10, 14]. Furthermore, the coseismic deformation of the Lushan earthquake is not consistent with the tectonics and geomorphology of the shallow sediments and as such there is an ongoing debate regarding the location, geometry, and kinematics of the Lushan seismogenic fault [22].

Presently, there are many uncertainties concerning the disastrous earthquakes in the LMS area [39]. For example, calculations of the Coulomb stress variations suggest that the previously accumulated stress was not completely released by the Wenchuan M_W 7.9 and Lushan M_W 6.7 earthquakes [6, 24]. Previous studies have warned that the LMS fault zone is active and presents a high risk [1] and thus, following a devastating seismic event, it is critical to determine where and when the next big earthquake will occur. Therefore, it is increasingly important to urgently survey active faults and assess potential seismic risks in the LMS area. Solving these problems is critical to understanding the seismogenic fault and seismotectonics of the Lushan earthquake and then to estimate the potential seismic hazard along the eastern margin of the Tibetan Plateau.

Legend: 1. Coseismic surface rupture of the Wenchuan earthquake; 2. Major faults; 3. Blind faults; 4. Aftershocks of the Wenchuan M_W 7.9 earthquake; 5. Aftershocks of the Lushan M_W 6.7 earthquake; F1. Longdong fault; F2. Wulong fault; F3. Xiaoguanzi fault; F4. Shuangshi-Dachuan fault; F5. Xinkaidian fault; F6. Dayi blind fault; F7. Sansuchang fault; SGB: Songpan–Ganze Block; LMS: Longmen Shan; SCB: Sichuan Basin; LQS: Longquan Shan.

5.2 Geological Setting

The approximately 500 km long and 30–50 km wide LMS thrust belt forms the eastern margin of the Tibetan Plateau (Fig. 5.1). The belt defines a sharp topographic transition between the Songpan–Ganze Block (SGB) and the Sichuan Basin (SCB) [2], which is a known seismically active area with complex intracontinental deformation [7, 15].

The LMS area has a long and complex tectonic history and its three main deformation stages since the Paleozoic have been identified [5, 19]. A number of extensional events occurred in the western SCB and LMS during the Paleozoic to Middle Triassic [23] that resulted in the formation of normal faults. Since the Mesozoic, the LMS thrust belt has had two major periods of contractional deformation, specifically in the late Triassic and the Cenozoic [13]. During the late Triassic, the tectonic setting changed from a rifted passive margin to a foreland in response to the final closure of the Paleo-Tethys and the continent–continent collisions of the North China, South China, and Qiangtang blocks [2]. During the Cenozoic, tectonism in the LMS was reactivated as a result of the India–Asia continental collision [30] and manifested in thrusting, dextral shear, uplifting, and exhumation [8].

The LMS area currently comprises a series of northeast-trending thrust faults [2, 19, 22]. The Wenchuan earthquake occurred in the central LMS and produced two large surface ruptures along the major faults (Fig. 5.1). The southern LMS thrust belt is distinguishable by the Longdong fault (F1), the Wulong fault (F2), the Xiaoguanzi fault (F3), and the Shuangshi-Dachuan fault (F4). The deformation in the western SCB is characterized by imbricated thrusts and fault-related folds (Fig. 5.2). The Lushan earthquake occurred in the southern LMS, but field investigations have not revealed any evidence of surface rupture due to this earthquake [16, 37].

The outcrops of the southern LMS are mainly Mesozoic strata (Fig. 5.2). This area is characterized by Baoxing complex rocks. Surface geological and drilling data indicate that the shallow sedimentary cover of the southern LMS contains mainly Sinian–Middle Triassic shallow marine clastic and carbonate rocks and late Triassic

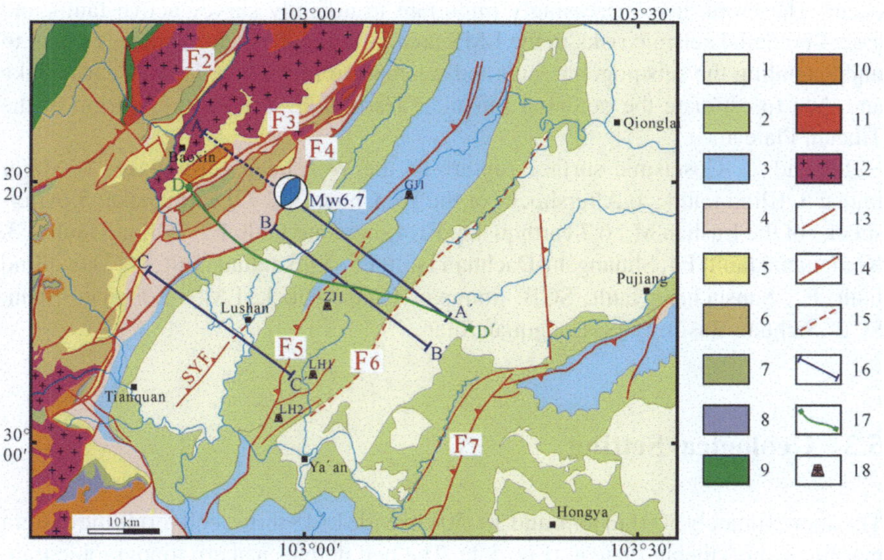

Fig. 5.2 Geology and location of artificial seismic reflection profiles in the research area. Legend: 1. Cenozoic; 2. Cretaceous; 3. Jurassic; 4. Triassic; 5. Permian; 6. Devonian; 7. Silurian; 8. Ordovician; 9. Cambrian; 10. Sinian; 11. Proterozoic; 12. Complex rocks; 13. Major faults; 14. Thrusts; 15. Blind faults; 16. Artificial seismic reflection profile; 17. Deep seismic reflection profile; 18. Drilling wells. SYF: Shiyang fault

non-marine clastics (Fig. 5.3). An early Proterozoic crystalline basement lies under this sedimentary cover.

There is an important shallow-level detachment (D1) that developed within the lower-Middle Triassic gypsum layers [13, 22]. This shallow detachment fault slips toward the foreland basin and forms the LQS anticline [20]. In addition, a potential deep-level detachment (D2) exists in the pre-Sinian crystalline basement beneath

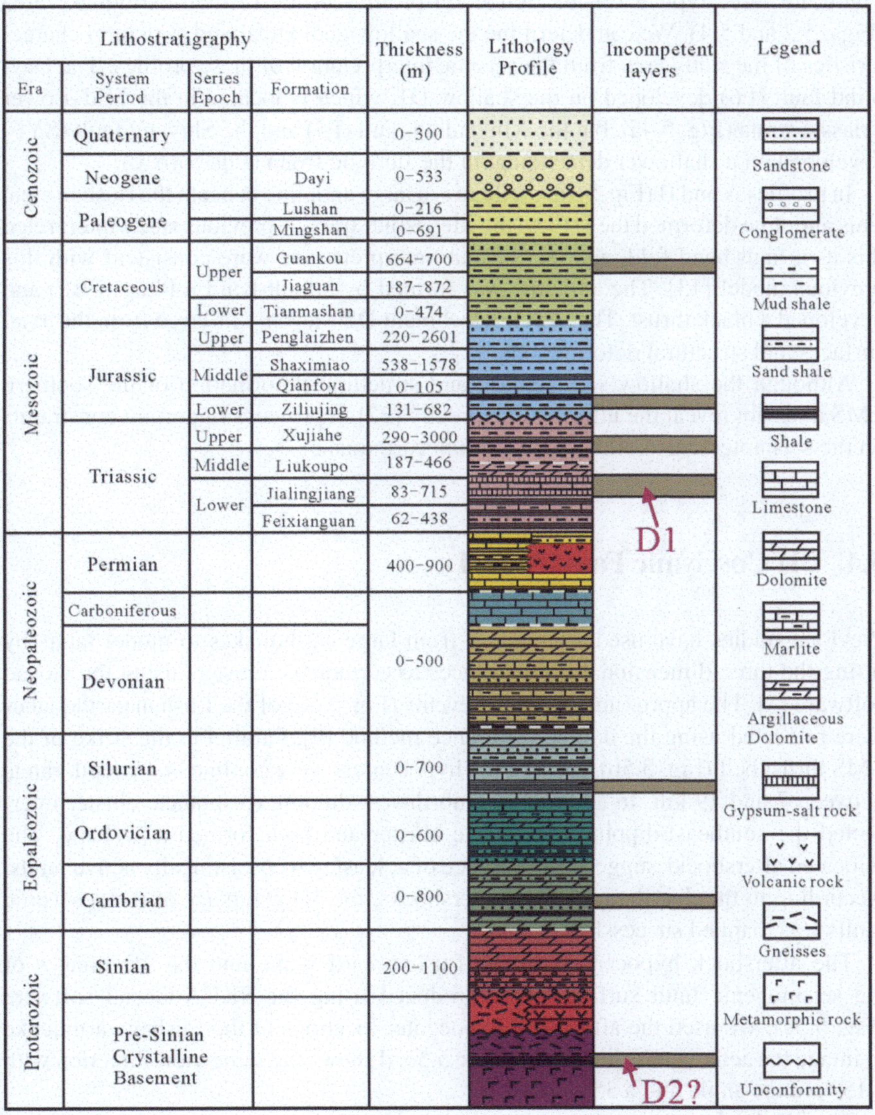

Fig. 5.3 Histogram of the regional stratigraphy of the LMS and the western SCB

the southeastern region of the Tibetan Plateau [31, 38]. Multiple detachments and superimposed deformation events are characteristic features of the southern LMS.

5.3 Shallow Structural Deformation

There are three typical seismic reflection profiles in the Lushan earthquake area (Figs. 5.2 and 5.4). We can determine the shallow geological and structural characteristics of the study area from the seismic interpretation of these profiles. The Dayi blind fault (F6) developed on the shallow D1, which is located in the Mid–Lower Triassic strata (Fig. 5.4a), but the Xinkaidian fault (F5) and the Shiyang fault (SYF) developed on a shallower detachment in the Jurassic strata (Fig. 5.4b, c).

In profiles A and B (Fig. 5.4a, b), there is a large anticline beneath the Lushan area. This anticline deformed the D1 and the Mesozoic strata. A previous study interpreted this as a fault-bend fold, and our seismic interpretations were consistent with this previous model [13]. The anticline was formed by a fault-bend folding (FBF) and developed a black thrust. The lower detachment D2 was only inferred from the axial surfaces and structural deformations.

Although the shallow sedimentary and structural deformation of the southern LMS does not reveal the information directly [22], it provides important constraints on the seismotectonics of the 2013 Lushan earthquake.

5.4 3D Coseismic Fault Model

Previous studies have used aftershocks from large earthquakes to model faults by fitting the three-dimensional (3D) surfaces to earthquake clusters using the Gocad software [3]. The approximately 2,000 events (Fig. 5.5a) of the Lushan earthquakes were relocated using the double-difference method [9]. Parallel to the strike of the LMS thrust belt (Fig. 5.5b), most of the hypocenters were confined to a depth range between 7 and 19 km. In addition, the northwest-dipping earthquake cluster intersected the southeast-dipping earthquake cluster and both formed a Y-shape. The relocated aftershocks suggest the presence of at least two coseismically active faults. According to the distribution of the aftershocks, the 3D geometry of the coseismic faults was mapped successfully [21].

The aftershock hypocenters of the Lushan earthquake and the 3D images of the seismogenic fault surfaces were produced using the SKUA-Gocad software (Fig. 5.5c). We used the aftershock hypocenter locations of the Lushan earthquake to image the active faults in detail. Figure 5.5c, d shows the same view direction with a 15° plunge angle and a 35° strike.

The 3D depth-contoured surfaces illustrate the detailed geometry of faults f1 and f2 (Fig. 5.5d). The surfaces of these faults are inhomogeneous and contain multiple

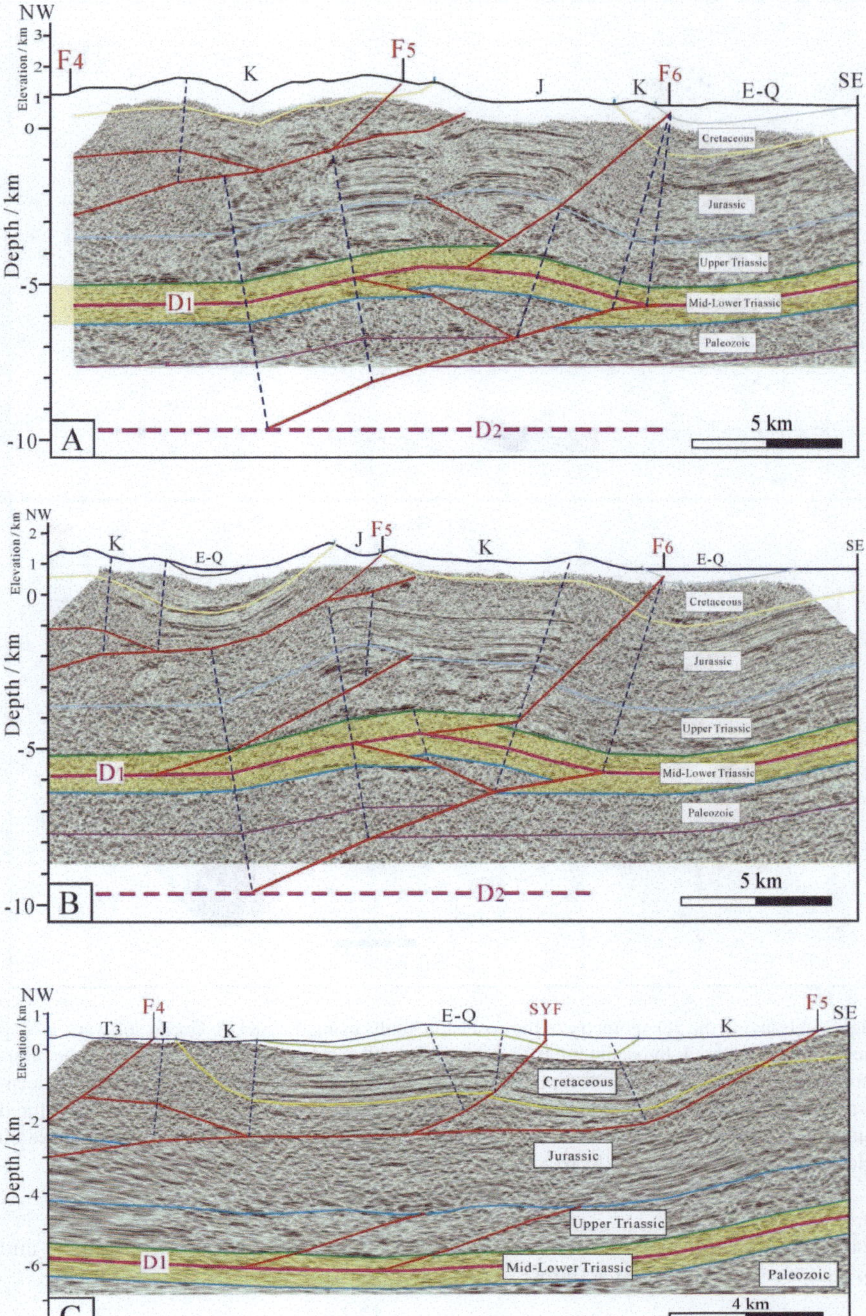

Fig. 5.4 Geological and structural interpretation of the seismic reflection profiles A, B, and C. The yellow area represents the Mid–Lower Triassic and detachment layer (D1). The red solid lines represent the faults and the black dashed lines represent the axial surfaces

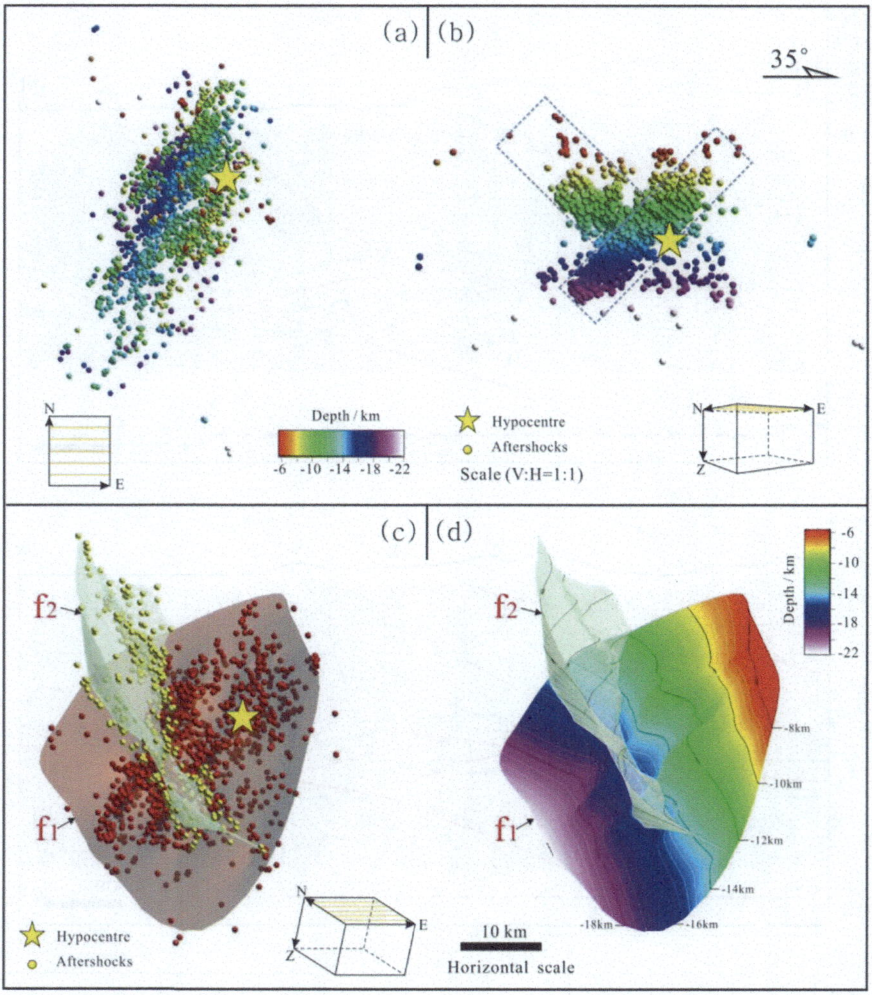

Fig. 5.5 Imaging the 3D geometry of the coseismic faults using the SKUA-Gocad software. **a** Map view of the aftershock hypocenters. **b** View parallel to the strike; the strike direction of the aftershocks is ~55°. **c** 3D images of the ruptured fault surfaces modeled based on the aftershock hypocenters after smoothing with the Gocad DSI algorithm. **d** Depth-contoured surfaces of the ruptured faults. The coordinate system is WGS 84 UTM 48 in the northern hemisphere. Approximately 2,000 relocated aftershocks were used ($M \geq 1.0$)

wave bends. Fault f1 is the major seismogenic fault of the Lushan earthquake and has a 45°–50° dip, whereas fault f2 is a branching backthrust of f1.

5.5 Seismic Interpretation and Quantitative Analysis

Profile E–E′ is a high-resolution petroleum industry seismic reflection profile between the southern LMS and western SCB (Fig. 5.6). This profile crosses the major fault-folding belt, and provides information on the shallow sediments and structures from the subsurface down to a depth of ~10 km. The Lushan syncline lies between the F4 and Xinkaidian fault (F5) (Figs. 5.1 and 5.2) and the hypocenter of the Lushan earthquake is beneath the F4 at a depth of 12–13 km [9, 41].

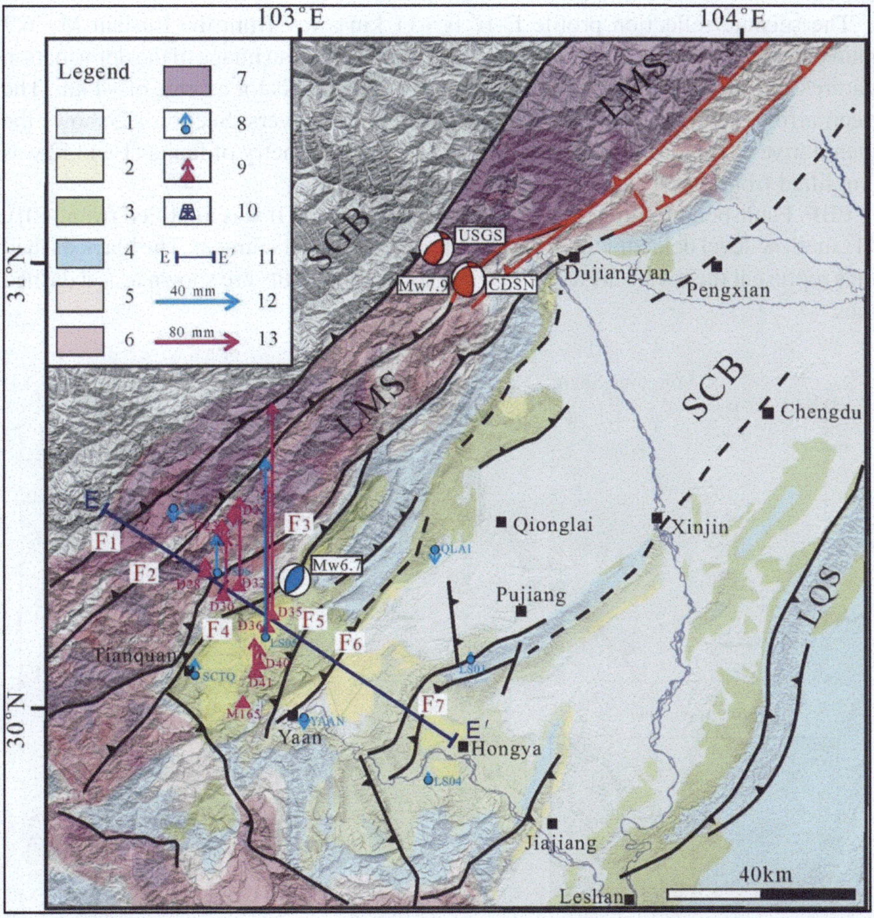

Fig. 5.6 Geological map of the southern LMS with the locations of the artificial seismic reflection profile, drilling wells, and GPS and leveling stations described in this study. Legend: 1. Quaternary; 2. Paleogene; 3. Cretaceous; 4. Jurassic; 5. Triassic; 6. Paleozoic; 7. Proterozoic and complex rocks; 8. GPS stations and measurements; 9. Leveling stations and measurements; 10. Drilling wells; 11. Petroleum industry seismic reflection profile; 12. Scale for GPS data [14]; and 13. Scale for leveling data [10]

In this seismic profile (Fig. 5.7), the labels on each layer denote certain seismic reflectors in the sediment rather than stratigraphic boundaries. The seismic reflectors in the Lushan syncline area are clear and continuous, suggesting that no thrust faults extend into the syncline; however, the partially enlarged portion of the seismic profile clearly shows the deformation features on top of the aftershock cluster (Fig. 5.7E1). The seismic reflectors of the Lower Paleozoic units are discontinuous and may suggest the presence of a fault that cuts across the Lower Paleozoic and some Upper Paleozoic layers [21]. The location of this fault is consistent with the location of the 3D-mapped seismogenic fault f1 (Fig. 5.5). Obviously, it has not broken through the Upper Triassic strata and is possibly a blind fault.

The seismic reflection profile E–E′ is ~11 km away from the Lushan M_w 6.7 mainshock. Section E1 displays a detailed seismic reflection image of the deformation features and strong seismic reflectors. It shows aftershocks for a range of ~4 km. The green arrows indicate the characteristics of the same layers. Section E2 shows the quantitative angle measurement of the fault f3. The geometry of faults f1 and f2 was simplified from the 3D images.

FBF: Fault-bend folding (the simple shear FBF model II was modified from [40]). D1: shallow-level detachment; D2: possible deep-level detachment. The black dashed lines represent the axial surfaces and some speculated faults are shown as red dashed lines.

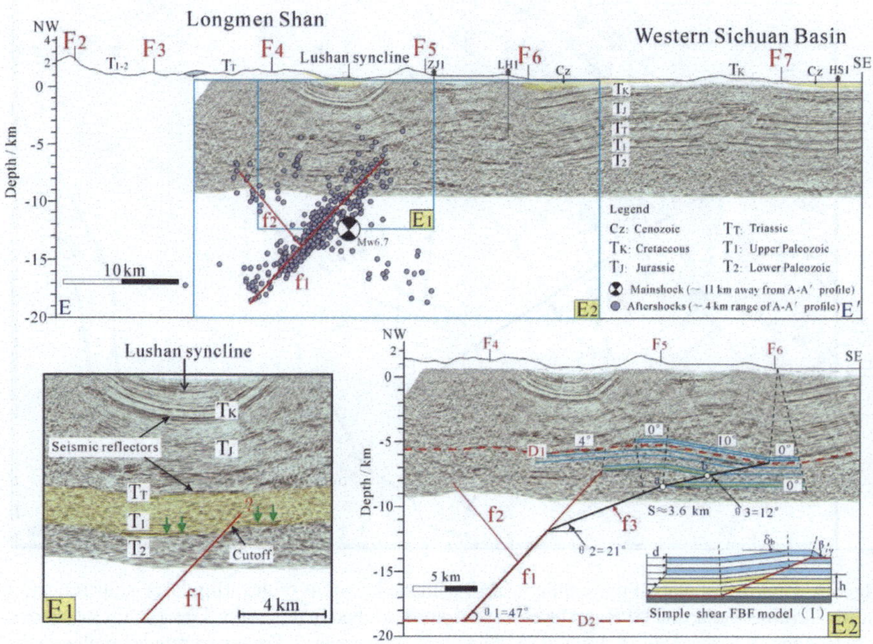

Fig. 5.7 Comprehensive analysis of the high-resolution seismic reflection profile E–E′

The detachments at the various levels may play a prominent role in the LMS and western SCB (Fig. 5.3). Some shallow detachments are parallel to the sedimentary layers and lead to shallow structural deformation [20]. Some detachments have propagated into the western SCB and produced fault-related folds during the Cenozoic, such as the Longquan Shan (LQS) anticline and Xiongpo anticline [12, 17, 33]. The seismic interpretation of profile E–E′ suggests that a shallow D1 occurs in the Mid–Lower Triassic gypsum beds. Some Mesozoic- and Cenozoic-age thrusts have developed on this detachment in the western SCB (Figs. 5.3 and 5.7E2). According to the S-wave velocity structure and structural analysis, a deep D2 is inferred to exist in the basement rocks at a depth of ~20 km [32]; however, the D2 may not be a regular horizontal surface at an accurate depth. Based on the distribution of the Lushan aftershocks and the seismic interpretation (Figs. 5.5 and 5.7), the D2 was determined to be located at a depth of approximately 19 km in this profile.

Between detachments D1 and D2, there is a large anticline beneath the F5 (Fig. 5.7), which is similar to the three shallow seismic profiles (Fig. 5.4). The f1 and f2 faults were simplified as two straight lines from the 3D images (Fig. 5.5). A main thrust fault f3 cuts the basement and merges with D1. We measured the dips of the layers and calculated the dip of fault f3 using the classical FBF theory [28]. The accumulated slip of fault f3 is ~3.6 km (Fig. 5.7E2). The seismic profile indicates that the f3 fault cuts the layers beneath D1 with a cutoff angle ($\theta 3$) of 12°. The cutoff angle $\theta 2$ is ~21° and fault f3 likely converges with fault f1 at a depth of ~12 km [21]. There is no evidence to show that the basal layer slides shorten and thicken above the ramp. The D2 detachment layer undergoes contraction and has a simple bedding-parallel surface with no basal fault slip but the layers above the ramp have no bedding-parallel surface. The geometry of fault f3 and the analysis of the axial surfaces suggest that the structure of f3 is consistent with the shear FBF model, which is characterized by long, gentle back limbs that dip less than the fault ramp [29].

Geodetic crustal displacement data is important for the study of active tectonics [27, 34], and can provide independent constraints for fault kinematics, particularly when no surface rupture features are available [14]. The GPS and leveling data collected during the 2013 Lushan earthquake indicate that the maximum coseismic deformations occurred in the Lushan syncline area (Fig. 5.8). In particular, at GPS station LS05 located ~15 km south of the epicenter an ~68 mm horizontal displacement and ~84 mm uplift were observed (Fig. 5.8E3). The angle of the coseismic deformation was approximately 50°, which is similar to the dip of fault f1. The leveling data suggest that the maximum coseismic displacements appeared at station D35 (Fig. 5.8E4) with an uplift of ~198.4 mm [10].

Section E3 shows the geological and structural interpretation of the seismic section in Fig. 5.7E2. The coseismic deformation obtained from the GPS data is shown at the top. The simple shear FBF model II was modified from [29]. Section E4 shows the geometry and kinematics model of seismogenic fault f1. The coseismic deformation of the leveling data is displayed above fault f1. Layer 1 indicates the deformation and the total uplift area, which are similar to the coseismic deformation of the leveling data.

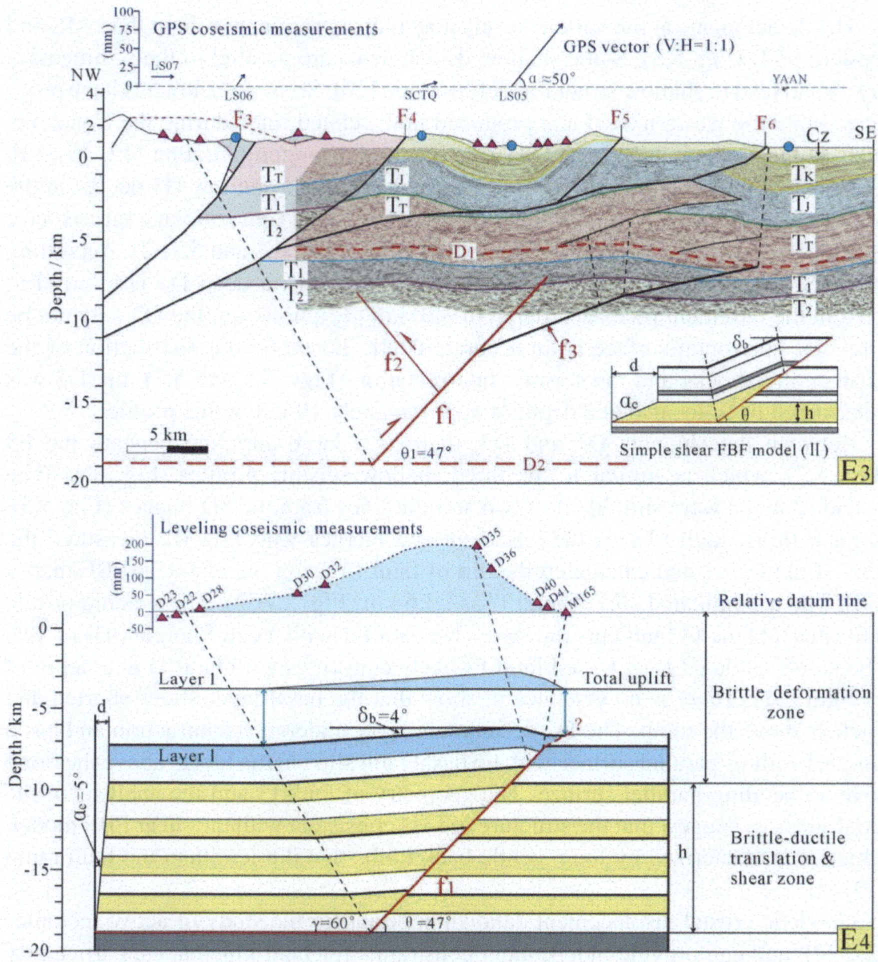

Fig. 5.8 Quantitative analysis of the geometry and kinematics of the seismotectonics of the 2013 Lushan earthquake

In this study, we mapped the area of the uplift resulting from the coseismic deformation and found that it has broad limb rotation in the back limb (Fig. 5.8E4). The back-dip (δb) measured from the seismic reflection profile is 4° and the ramp dip (θ1) is 47°. According to the simple shear FBF model and relevant formulae [26, 29], the shear angle (αe) is ~5°. The instantaneous kinematics and total uplift of the coseismic deformations indicate that the seismotectonics of the Lushan earthquake correlate with the simple shear FBF model [40], which successfully explains the back limb uplift [21]. This result suggests that the upper crust in the study area has a brittle deformation zone above ~10 km and a brittle–ductile translation or shear zone between 10 and 20 km. This brittle–ductile translation zone is consistent with the results of previous studies [32, 38].

5.6 Seismotectonics of the Lushan Earthquake

A deep seismic reflection profile D–D′ was collected following the 2013 Lushan earthquake (Fig. 5.2). This seismic profile shows the image of a deep structure beneath the Lushan syncline (Fig. 5.9). The seismic reflectors suggest that the Triassic and some Upper Paleozoic layers are roughly continuous [21], which is consistent with the other profiles (Figs. 5.4 and 5.7). However, the Lower Paleozoic layers exhibit a prominent offset above the mainshock of the Lushan earthquake, which suggests the presence of a fault cutting the Lower Paleozoic and basement rocks.

The focal mechanism of the 2013 Lushan earthquake suggests that the seismogenic fault f1 is a thrust fault (Fig. 5.1). However, the deep seismic profile D–D′ shows that it is a typical normal fault, as there are thin layers in the footwall of f1 but thick sediments in the hanging wall between the Upper Paleozoic and Lower Paleozoic units (Fig. 5.9). These syndepositional features formed in response to the growth of fault f1. Based on the tectonic evolution of the LMS [19, 23], we suggest that fault f1 was a normal fault during the Paleozoic (Fig. 5.9D1).

The seismogenic fault f1 involves positive inversion tectonics. It is just starting to form (Fig. 5.9D2); however, as it has not broken through the postrift strata, as

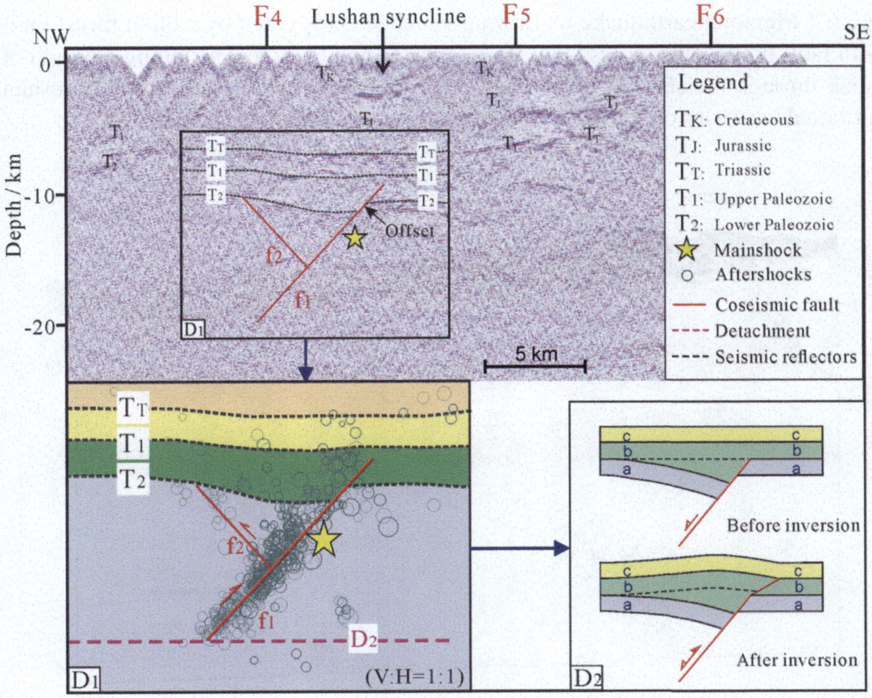

Fig. 5.9 Deep artificial seismic reflection profile D–D′ and the coseismic faults of the Lushan earthquake

positive inversion structural evolution models suggest [35], and has barely caused any structural deformations. Fault f3 is the main fault with a slip at ~3.6 km (Fig. 5.7E2), which suggests that it formed and tectonically inverted during the Late Cenozoic.

In this figure, the seismically active f1 and f2 faults are simplified as two straight lines. Section D1 shows the coseismic faults and the geological model of the 2013 Lushan earthquake, suggesting that fault f1 was a normal fault prior to the Mesozoic period. Section D2 shows the schematic of a classical positive inversion structure. The stratigraphic sequences "a", "b", and "c" are prerift, synrift, and postrift, respectively (modified from [35]).

Previous studies demonstrated the formation of several normal faults prior to the Late Triassic period in the LMS area [2, 13]. Since the Cenozoic, some pre-existing syndepositional normal faults facilitated the structural inversion that transformed them into thrust faults in a compressional regime [5]. In the LMS area and the western SCB, several of these pre-existing normal faults were structurally inverted and transformed into active thrust faults in a contractional setting [7, 30].

Clearly, the seismotectonics of the 2013 Lushan M_W 6.7 earthquake are different from the seismotectonics of the 2008 Wenchuan M_W 7.9 earthquake (Fig. 5.10). The Wenchuan earthquake ruptured the upper crust with two main faults in the LMS [36], but the seismogenic fault f1 responsible for the 2013 Lushan earthquake is a blind fault that was reactivated during the Late Cenozoic [22]. It also differs from the 2016 M_W 6.4 Meinong earthquake in Taiwan, which was triggered by a blind thrust fault with fault propagation folds [11]. Both faults f1 and f3 were developing but did not break through the shallow detachment D1. Therefore, we suggest that the Lushan earthquake activity on the seismogenic fault f1 is a hidden earthquake [25].

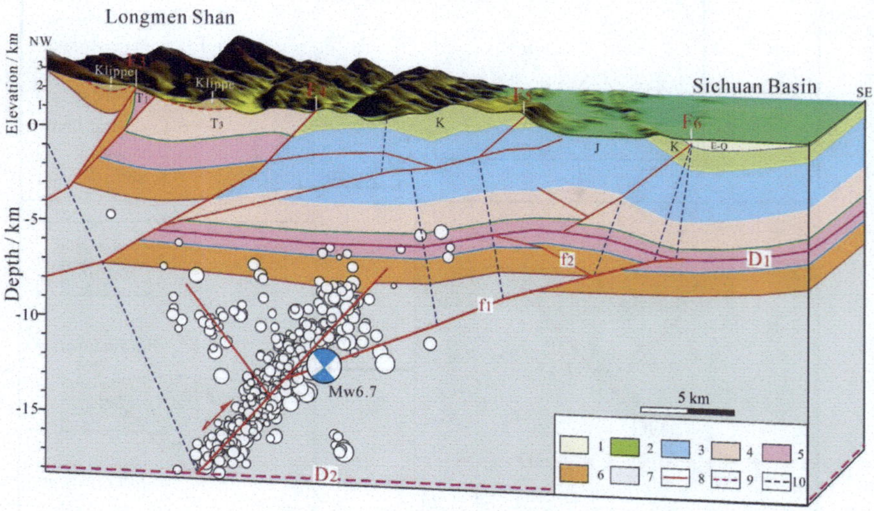

Fig. 5.10 A 3D geological and structural model showing seismotectonics of the Lushan earthquake. Legend: 1. Cenozoic; 2. Cretaceous; 3. Jurassic; 4. Upper Triassic; 5. Mid–Lower Triassic; 6. Upper Paleozoic; 7. Pre-Cambrian and basement; 8. Major faults; 9. Main detachments; 10. Axial surfaces

It is known that many active and blind faults exist in the frontal LMS and western SCB [33]; furthermore, several scientists believe that the accumulation of strain energy in the southern LMS area was not completely released by the 2013 Lushan event [6]. The seismotectonics of the 2013 Lushan earthquake show that inversion tectonics could trigger a strong earthquake, and these were an important factor in the development of the southern LMS during the Late Cenozoic. Furthermore, the blind fault led to a hidden earthquake, which suggests that the frontal ramp of these reactivated faults poses a potential seismic risk. With the ongoing compression of the Tibetan Plateau toward the western SCB during the Late Cenozoic [42], the level of seismic risk in the LMS and western SCB area is increasingly uncertain.

5.7 Conclusions

The 2013 M_W 6.7 event was a hidden earthquake that was ruptured by an unknown blind thrust fault in the frontal Longmen Shan (LMS). Three-dimensional imaging of the aftershocks was used to identify two planar faults that form a Y-shape with the major seismogenic fault f1 dipping to the northwest at 45°–50° at a depth of 7–19 km. Seismic interpretations suggest that the seismogenic fault is a typical blind fault that did not penetrate into the overlying Mesozoic and Cenozoic units and it is not a Shuangshi-Dachuan fault (F4) or a frontal Dayi buried fault (F6). Deep seismic data illustrate the syndepositional characteristics of the seismogenic fault. Our study further demonstrated that the seismogenic slip occurred on a reactivated pre-existing normal fault older than the Triassic age that triggered earthquakes during the Late Cenozoic.

There are two major detachments that control the structural deformation of the upper crust in the LMS and western Sichuan Basin (SCB), resulting in multiple superimposed deformation events. Based on the structural and coseismic deformation, we argue that the main seismotectonics of the Lushan earthquake are consistent with the simple shear fault-bend folding model. Clearly, the 2013 Lushan earthquake was not a characteristic earthquake. Due to the strong compression of the Tibetan Plateau toward the SCB, the early-period normal faults were activated after inversion and triggered the Lushan earthquake. The concealed and reactivated faults increase the potential risks related to earthquakes in the LMS and the western SCB.

Acknowledgements This study receives financial support from the National Key R&D Program of China (grant 2018YFC1504104), the National Natural Science Foundation of China (41872206). The body-wave moment tensor solution database was from the Institute of Geophysics, China Earthquake Administration (IGCEA) (http://www.cea-igp.ac.cn/tpxw/266824.html, last accessed April 2013). Information on the 2013 Lushan aftershocks was last downloaded from the China Earthquake Networks Center (CENC) (http://www.ceic.ac.cn/, last accessed April 2013). We would like to thank the Data Management Centre of the China National Seismic Network for their assistance and Petro China for providing the artificial seismic reflection profiles and drilling well data.

References

1. Broadbent N (2014) Longman shan fault zone still hazardous, suggest new reports. Seismol Res Lett http://www.eurekalert.org/pub_releases/2014-01/ssoa-lfz122513.php
2. Burchfiel BC, Chen Z, Liu Y, Royden LH (1995) Tectonics of the Longmen Shan and adjacent regions, central China. Int Geol Rev 37:661–735. https://doi.org/10.1080/00206819509465424
3. Carena S, Suppe J (2002) Three-dimensional imaging of active structures using earthquake aftershocks: the Northridge thrust, California. J Struct Geol 24:887–904
4. Chen LC, Wang H, Ran YK, Lei SK, Li X, Wu FY, Ma XQ, Liu CL, Han F (2014) The 2013 Lushan Ms 7.0 earthquake: varied seismogenic structure from the 2008 Wenchuan earthquake. Seismol Res Lett 85:34–39. https://doi.org/10.1785/0220130109
5. Chen SF, Wilson CJL (1996) Emplacement of the Longmen Shan thrust-nappe belt along the eastern margin of the Tibetan Plateau. J Struct Geol 18:413–430
6. Chen YT, Yang ZX, Zhang Y, Liu C (2013) From 2008 Wenchuan earthquake to 2013 Lushan earthquake. Scientia Sinica Terrae 43:1064–1072
7. Deng QD, Chen SF, Zhao XL et al (1994) Tectonics, seismicity and dynamics of Longmen Shan mountains and its adjacent regions (in Chinese). Seismol Geol 16(4):389–403
8. Densmore AL, Ellis MA, Li Y, Zhou RJ, Hancock GS, Richardson N (2007) Active tectonics of the Beichuan and Pengguan faults at the eastern margin of the Tibetan Plateau. Tectonics 26, TC4005
9. Fang LH, Wu JP, Wang WL, Du WK, Su JR, Wang CZ, Yang T, Cai Y (2015) Aftershock observation and analysis of the 2013 Ms 7.0 Lushan earthquake. Seismol Res Lett 86:1–8. https://doi.org/10.1785/0220140186
10. Hao M, Wang QL, Liu LW, Shi Q (2014) Interseismic and coseismic displacements of the Lushan Ms7.0 earthquake inferred from leveling measurements. Chin Sci Bull 59:5129–5135. https://doi.org/10.1007/s11434-014-0652-4
11. Huang MH, Tung H, Fielding E, Huang HH, Cunren L, Chuang H, Hu JC (2016) Multiple fault slip triggered above the 2016 Mw 6.4 MeiNong earthquake in Taiwan. Geophys Res Lett. https://doi.org/10.1002/2016gl069351
12. Hubbard J, Shaw JH (2009) Uplift of the LMS and Tibetan Plateau, and the 2008 Wenchuan (M = 7.9) earthquake. Nature 458:194–197. https://doi.org/10.1038/nature07837
13. Jia D, Wei GQ, Chen ZX, Li BL, Zeng Q, Yang G (2006) Longmen Shan fold-thrust belt and its relation to western Sichuan Basin in central China: new insights from hydrocarbon exploration. Am Assoc Petrol Geol 90:1425–1447. https://doi.org/10.1306/03230605076
14. Jiang ZS, Wang M, Wang YZ, Wu YQ, Che S, Shen ZK, Bürgmann R, Sun JB, Yang YL, Liao H, Li Q (2014) GPS constrained coseismic source and slip distribution of the 2013 Mw6.6 Lushan, China, earthquake and its tectonic implications. Geophys Res Lett 41:407–413
15. Kirby E, Whipple K, Harkins N (2008) Topography reveals seismic hazard. Nat Geosci 1:485–487. https://doi.org/10.1038/ngeo265
16. Lei SX, Ran YK, Wang H et al (2014) Discussion on whether there are coseismic surface ruptures of the Lushan Ms7.0 earthquake at Longmen area and its implications (in Chinese). Seismol Geol 36(1):266–274
17. Li Y, Jia D, Wang M, Shaw JH, He J, Lin A, Xiong L, Rao G (2014) Structural geometry of the source region for the 2013 Mw 6.6 Lushan earthquake: implication for earthquake hazard assessment along the Longmen Shan. Earth Planet Sci Lett 390:275–286. https://doi.org/10.1016/j.epsl.2014.01.018
18. Li Y, Zhou RJ, Zhao GH et al (2013) Thrusting and detachment folding of Lushan earthquake in front of Longmenshan Mountains (in Chinese). J Chendu Univ Technol 40(4):353–363
19. Liu SG (1993) The formation and evolution of Longmengshan thrust zone and Western Sichuan, China (in Chinese). Press of Chengdu University of Science and Technology, Chengdu, pp 17–117
20. Lu RQ, He DF, Xu XW et al (2016) The basement structures beneath the Central segment of Longmen ShanShan: constraints on the uplifting in the southeast margin of Qinghai-Tibet Plateau since the Cenozoic. J Asian Earth Sci 117:73–81

21. Lu RQ, Xu XW, He DF et al (2017) Seismotectonics of the 2013 Lushan Mw 6.7 earthquake: inversion tectonics in the eastern margin of the Tibetan Plateau. Geophys Res Lett 44:8236–8243
22. Lu RQ, Xu XW, He DF et al (2017) The shallow sedimentary and structural deformation in the southern Longmen Shan: constraints on the seismotectonics of the 2013 Lushan Mw6.7 earthquake. Chin J Geophys 60(8):2924–2934
23. Luo ZL, Jin YZ, Zhu KY, Zhao XK (1988) On Emei Taphrogenesis of the upper Yangze platform. Geol Rev 34:11–23
24. Parsons T, Ji C, Kirby E (2008) Stress changes from the 2008 Wenchuan earthquake and increased hazard in the Sichuan basin. Nature 454:509–510. https://doi.org/10.1038/nature07177
25. Stein RS, Yeats RS (1989) Hidden—earthquakes. Sci Am 260:48–57
26. Shaw JH, Connors C, Suppe J (2004) Seismic interpretation of contractional fault-related folds: an american association of petroleum geologists seismic atlas. Am Assoc Petrol Geol Spec Publ
27. Shen Zheng-Kang, Jianbao Sun, Zhang Peizheng (2009) Slip maxima at fault junctions and rupturing of barriers during the 12 May 2008 Wenchuan earthquake. Nat Geosci 2:718–724
28. Suppe J (1983) Geometry and kinematics of fault-bend folding. Am J Sci 283:684–721
29. Suppe J, Connors CD, Zhang YK (2004) Shear fault bend folding. In McClay KR (ed) Thrust tectonics and hydrocarbon systems, AAPG Memoir, vol 82, pp 303–323
30. Tapponnier P, Xu Z, Roger F, Meyer B, Arnaud N, Wittlinger G, Yang J (2001) Oblique stepwise rise and growth of the Tibet Plateau. Science 294:1671–1677. https://doi.org/10.1126/science.105978
31. Tang LJ, Yang KM, Jin WZ et al (2008) Multi-detachment and the structural deformation of Longmen mountain thrust belt. Sci China (in Chinese) 38:30–40
32. Wang CY, Lou H, Silver PG et al (2010) Crustal structure variation along 30° N in the eastern Tibetan Plateau and its tectonic implications. Earth Planet Sci Lett 289:367–376
33. Wang M, Jia D, Shaw JH, Hubbard J, Plesch A, Li Y, Liu BJ (2014). The 2013 Lushan earthquake: implications for seismic hazards posed by the range front blind thrust in the Sichuan Basin, China. Geology 42:915–918. https://doi.org/10.1130/G35809.1
34. Wang Qi, Qiao Xuejun, Lan Qigui (2011) Rupture of deep faults in the 2008 Wenchuan earthquake and uplift of the Longmen Shan. Nat Geosci 4:633–640
35. Williams GD, Powell CM, Cooper MA (1989) Geometry and kinematics of inversion tectonics. Geol Soc 44:3–15. https://doi.org/10.1144/GSL.SP.1989.044.01.02
36. Xu XW, Wen XZ, Yu GH, Chen GH, Klinger Y, Hubbard J, Shaw J (2009) Co-seismic reverse- and oblique-slip surface faulting generated by the 2008 Mw 7.9 Wenchuan earthquake, China. Geology 37(6):515–518. https://doi.org/10.1130/g25462a.1
37. Xu XW, Wen XZ, Han Z et al (2013) Lushan Ms7.0 earthquake: a blind reserve-fault earthquake (in Chinese). Chin Sci Bull 58. https://doi.org/10.1007/s11434-013-5999-4
38. Yan DP, Zhou MF, Li SB, Wei GQ (2011) Structural and geochronological constraints on the Mesozoic-Cenozoic tectonic evolution of the Longmen Shan thrust belt, eastern Tibetan Plateau. Tectonics 30:TC6005. https://doi.org/10.1029/2011tc002867
39. Yin A (2010) Preface: a special issue on the great 12 May 2008 Wenchuan earthquake (Mw7.9): observations and unanswered questions. Tectonophysics 491:1–9
40. Yue LF, Suppe J, Hung JH (2011) Two contrasting kinematic styles of active folding above thrust ramps, western Taiwan. In: McClay K, Shaw J, Suppe J (eds) Thrust fault-related folding. AAPG Memoir 94:153–186
41. Zeng XF, Luo Y, Han LB, Shi YL (2013) Lushan Ms7.0 earthquake on 20 April 2013: a high-angle thrust event. Chin J Geophys 56:1418–1421. https://doi.org/10.6038/cjg20130437
42. Zhang P, Shen Z, Wang M, Gan W, Burgmann R, Molnar P (2004) Continuous deformation of the Tibetan Plateau from global positioning system data. Geology 32:809–812

Chapter 6
Earthquake Focal Mechanisms and Stress Environment in Southern Taiwan

Li Zhao, Chiao-Ying Lee, and Shu-Huei Hung

Abstract We conduct a systematic examination of the stress field environment surrounding the 4 March 2010 Jiahsian earthquake (M_W 6.3) in southern Taiwan. First, we perform waveform inversion to determine the focal mechanisms of 318 earthquakes of magnitudes $3.5 \leq M_L \leq 6.4$ that occurred between January 1997 and December 2010 within 50 km of the Jiahsian epicenter. These focal mechanisms are then used to derive the orientations of the principle stress axes using a damped regional-scale stress inversion algorithm. Our results show that southern Taiwan is dominated by two types of stress environments: the overall northwest–southeast compression in the Western Foothills and the Longitudinal Valley resulting from the convergence of the Philippine Sea and Eurasian plates and the northeast–southwest extension due to the exhumation of the Central Range. Moderation by local conditions led to areas with distinct types of stress conditions: thrust in the Western Foothills and the Longitudinal Valley, strike-slip in the Central Range north of 23° N, normal in the southeastern Central Range, and mixed-type across the Chaochou Fault. Significant earthquakes, such as the 1999 Chi-Chi and the 2010 Jiahsian earthquakes, have a detectable influence on the regional stress field.

Keywords Focal mechanism · Stress field · Chi-Chi earthquake · Jiahsian earthquake · Taiwan

6.1 Introduction

Taiwan is situated in the collision zone between the Eurasian and the Philippine Sea Plates. The Philippine Sea Plate (PSP) moves northwestward at a rate of 82 mm/year [21] relative to the Eurasian Plate (EP). Off the eastern coast of Taiwan, the PSP

L. Zhao (✉)
School of Earth and Space Sciences, Peking University, Beijing, China
e-mail: lizhaopku@pku.edu.cn

C.-Y. Lee · S.-H. Hung
Department of Geosciences, National Taiwan University, Taipei, Taiwan

C.-H. Lo et al. (eds.), *Earthquake Geology and Tectonophysics around Eastern Tibet and Taiwan*, Atmosphere, Earth, Ocean & Space,
https://doi.org/10.1007/978-981-15-6210-5_6

submerges northward under the EP, whereas in the south of the Taiwan Island, the EP plunges eastward beneath the PSP. In between these two opposite subduction zones, along the southeastern coast of Taiwan, the PSP collides with the EP (see Fig. 6.1). This oblique collision leads to complicated spatial variation in the stress field as well as a high level of seismic activity in southern Taiwan. Previous studies on the regional stress field of Taiwan took advantage of the large number of available focal mechanisms of local earthquakes. Angelier et al. [3] investigated the paleo-stress field in Taiwan's Western Foothills (WF) and found that the whole region is in a compressional stress environment, with the maximum compressional axis

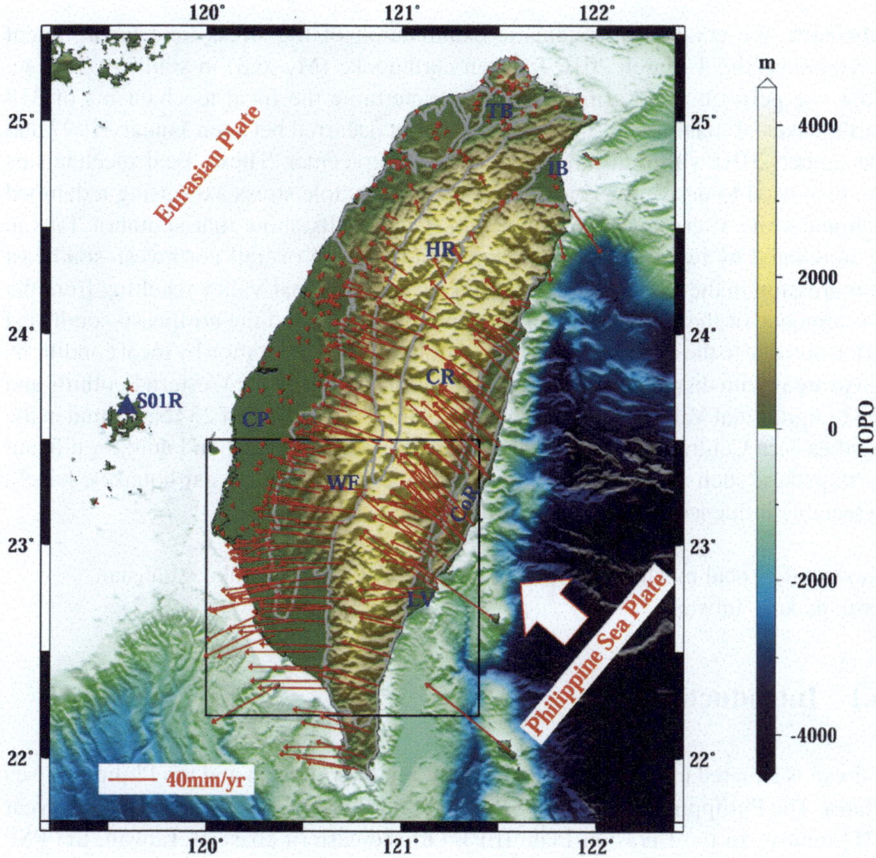

Fig. 6.1 Horizontal GPS velocity field on Taiwan from 1994 to 2011 relative to station S01R (blue triangle) on Penghu Island (*Source* http://gps.earth.sinica.edu.tw/). The large white arrow shows the average direction of convergence of the Philippine Sea Plate with the Eurasian plate, whereas the red arrows are GPS-measured horizontal velocities. The background color shows the topography. Gray lines show the geological boundaries. The geological blocks are denoted as follows: CP: coastal plain; WF: Western Foothills; HR: Hsueshan range; CR: Central Range; LV: Longitudinal Valley; CoR: Coastal range; TB: Taipei basin; and IB: Ilan basin

more or less parallel to the direction of the convergence between the PSP and EP. From north to south, the maximum compressional axis in the WF experiences a counter-clockwise rotation.

Yeh et al. [20] divided the island of Taiwan into 21 regions based on the focal mechanisms of 200 earthquakes and inverted the stress field for each region independently using Angelier's technique [1, 2], which minimizes the least-squares difference between the slip vectors of the earthquakes and the tangential tractions of the regional average stress field. The results also showed a largely horizontal maximum compressional axis in western Taiwan-oriented northwest–southeast, parallel to the relative motion between the PSP and EP.

Rau and Wu [14] determined the focal mechanisms of 97 earthquakes based on their P-wave polarities and the spectral amplitude ratios between S and P waves, and found that the earthquakes are mostly thrust in the western fold and thrust belt, but strike-slip and normal in the Central Range (CR). They also inverted the focal mechanisms for the orientations of the principle stress axes based on the technique of Gephart and Forsyth [6]. They found again that the maximum compressional axis is largely parallel to the plate convergence, and that the stress field is not uniform.

Wu et al. [19] presented a comprehensive study that systematically determined the focal mechanisms of shallow-focus earthquakes (30 km and above) by applying a genetic algorithm to P-wave first-motion polarization data, and then inverting for the stress field based on the method developed by Michael [12, 13]. Their result agreed with that of Yeh et al. [20], that is, in western Taiwan the horizontal maximum compressional axis is aligned with the direction of plate convergence. In the CR, at around 23 °N and 121 °E, they also found a normal faulting stress field with a nearly vertical maximum compressional axis and a nearly horizontal maximum extensional axis that is oriented northeast–southwest. In the eastern Longitudinal Valley (LV), the maximum compressional axis has an azimuth of approximately 115°, sub-parallel to the direction of plate convergence. Based on the same focal mechanism solutions, Hsu et al. [8] used Michael's method [12, 13] and a moving window approach to investigate the spatial variation of the stress field. They also found a normal faulting stress environment in the CR, which they interpreted as being induced by the exhumation of buried crustal material and crustal thickening. In addition, in the southern CR, the maximum compressional axis changes its orientation to the southwest, as a result of southwestern Taiwan's escaping tectonics. Hsu et al. [8] also inverted for the strain field using GPS observations. The principle strain axes are largely in line with the principle stress axes, indicating a depth-independent stress field.

Wu et al. [18] examined both the spatial and temporal variations of the stress field, and concluded that spatially the principle stress axes follow the tectonic divisions, whereas temporally they are influenced by the 1999 Chi-Chi earthquake, with the maximum compressional axis gradually returning to its orientation prior to the earthquake, which is consistent with the pattern observed from seismicity [17].

In this study, we focus on the stress field of southern Taiwan, in particular the region surrounding the 4 March 2010 Jiahsian earthquake, which had a magnitude of M_W 6.3 and represents the first major earthquake on the island of Taiwan after the 1999 Chi-Chi earthquake. We first determine the focal mechanisms of moderate and small earthquakes surrounding the Jiahsian epicenter, and then invert for the regional stress field. We examine the spatial and temporal changes in the stress field and discuss the variations of the tectonic stress field in southern Taiwan as well as its possible implications on the regional seismicity.

6.2 Seismic Data

We use the waveform records from the Broadband Array in Taiwan for Seismology (BATS; Fig. 6.2a) operated by the Institute of Earth Sciences, Academia Sinica, in Taiwan [9]. The selected earthquakes are of moderate and small magnitudes ($M_L \geq 3.5$) located within 50 km of the Jiahsian epicenter (22.97 °N, 120.71 °E), and occurred between January 1997 and December 2010. We retrieved the waveform records for a total of 694 events from the BATS data management center (http://tecdc.earth.sinica.edu.tw/). The distribution of earthquakes is shown in Fig. 6.2b. After initial examination based on the signal-to-noise ratio as well as event distribution, only 393 earthquakes contained waveforms of sufficiently high quality and were selected in subsequent focal mechanism determination. These included all events of magnitude $M_L \geq 4.0$ (252 out of 694) and 141 of magnitude $3.5 \leq M_L < 4.0$.

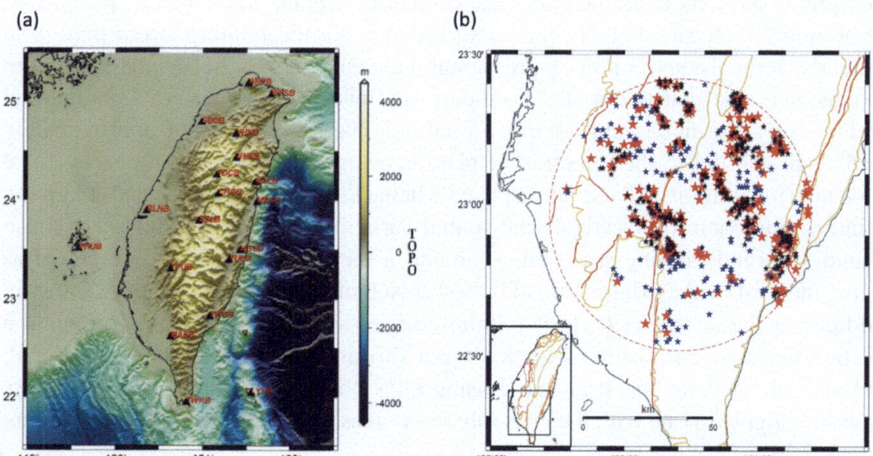

Fig. 6.2 **a** Distribution of seismic stations used in this study. Black triangles depict station locations. Background color indicates the topography. **b** Distribution of the 694 earthquakes of $3.5 \leq M_L \leq 6.4$ that occurred between January 1997 and December 2010 within 50 km of the epicenter of the 4 March 2010 Jiahsian earthquake. Red and blue stars represent the 393 earthquakes used in this study, among which the red stars represent the 252 earthquakes of $M_L \geq 4.0$, whereas the blue stars represent the 141 earthquakes of $3.5 \leq M_L < 4.0$. The black stars depict the 301 earthquakes of $3.5 \leq M_L < 4.0$ that are not used because of the lack of quality waveform records. Red lines are active faults, while orange lines are geological boundaries

6.3 Earthquake Focal Mechanism Determination

In this study, we adopt the cut and paste (CAP) method [22, 23] to determine the earthquake focal mechanisms. The CAP method uses a grid search to find the optimal set of earthquake focal parameters, including the strike, dip, and rake angles, the source depth, and the scalar moment, that yield the best fit between the records and synthetics. To maximize the usage of the seismic signals contained in the waveforms, three-component records at each station are cut into five separate windows: P-wave windows in the vertical and radial components, and surface-wave windows in vertical, radial, and transverse components. In the CAP method, the synthetics are computed using the frequency-wavenumber integration approach [24] in one-dimensional (1D) velocity models. To minimize the effect of lateral velocity heterogeneities, we used waveforms of relatively long period, and a time shift was allowed in the calculation of each waveform residual. The final waveform fitting was measured using a carefully defined error function that sums up the waveform residuals in all the windows with appropriate weights, taking into consideration the amplitude contrast between P waves and surface waves, and the attenuation of amplitude with epicentral distance. Since the CAP method uses the P-wave window that also contains the depth phases such as pP and sP, and the waveform fitting is independent of surface wave windows, it provides a more robust constraint on the earthquake source depth in addition to its mechanism.

In calculating the synthetics in this study, we used a 1D model obtained by averaging the tomographic model of Taiwan from Wu et al. [16]. The model contains P- and S-wave velocities from a joint inversion of P- and S-wave arrival times of hundreds of thousands of earthquakes in the Taiwan region. In the CAP inversion of focal mechanisms, the P waves and surface waves were bandpass filtered to the frequency bands of 0.05–0.3 Hz and 0.02–0.1 Hz, respectively, to balance the need for shorter-period body waves and the fitness of the 1D model. Figure 6.3 shows an example of the waveform fitting for an earthquake of magnitude M_W 3.76. The recorded and synthetic waveforms correlate well for most windows, ensuring reliable results for the source depth and focal mechanism.

Using the records of the 393 earthquakes, we obtained reliable focal mechanisms for 318 earthquakes, as displayed in Fig. 6.4. A clear regional pattern can be seen in the earthquake focal mechanisms in southern Taiwan, as indicated by the ellipses A, B, C, D, and E in Fig. 6.4. Earthquakes in region A are mostly shallow-focus thrust events, with nearly horizontal P-axes oriented in a northwest–southeast direction. In region B in the CR, the earthquakes are still shallow, but the mechanisms have become strike-slip. Further south in the CR, the earthquakes in region D are again shallow but turn to normal events with northeast–southwest trending T-axes. Events in the southern LV in region C have a mixture of thrust and strike-slip mechanisms, with P-axes consistently oriented in a northwest–southeast direction, parallel to the plate convergence, and T-axes with no prevailing orientation. Finally, in region E toward the southern end of the CR, the earthquakes are deeper with diverse mechanisms.

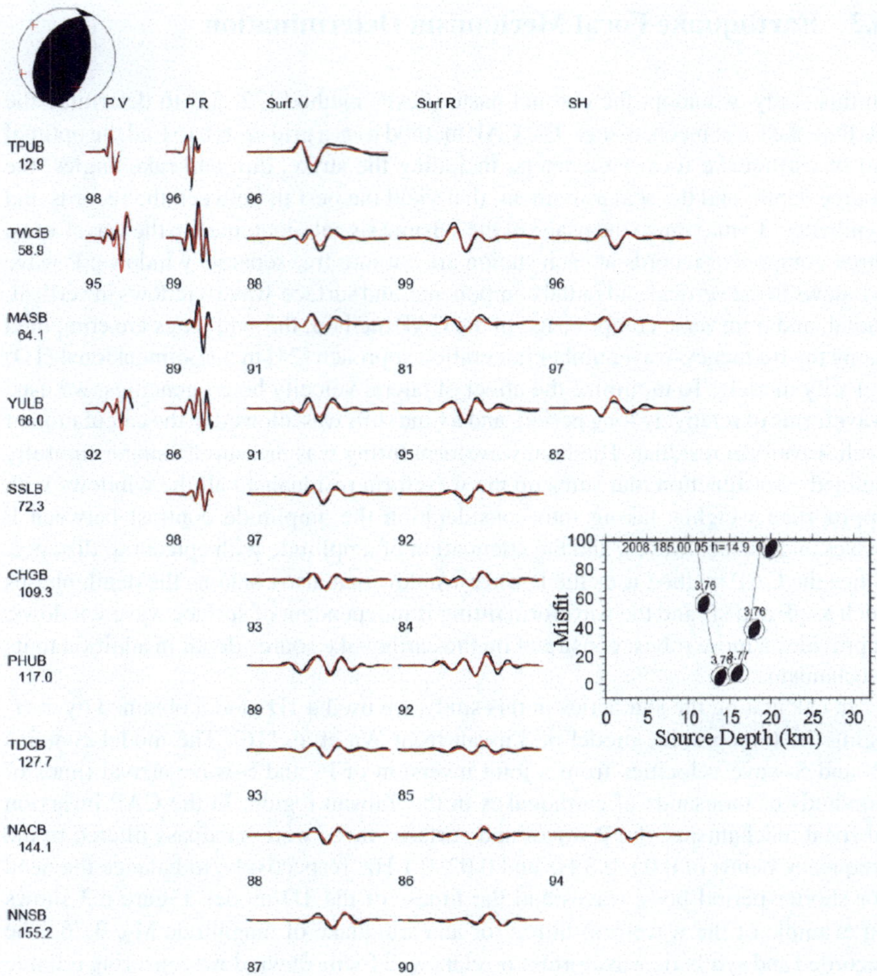

Fig. 6.3 Example of the CAP solution for an earthquake. The final solution is displayed by the beachball in the top-left corner. Records (black curves) and synthetics (red curves) are compared at ten stations. On each row of the waveform plots, the station name and epicentral distance (in kilometers) are given on the left, followed by the waveform comparisons in five windows: vertical- and radial-component P waves; and vertical-, radial-, and transverse-component surface waves. The number under each waveform is the record-synthetic cross-correlation coefficient (in percentage). The plot on the right shows the result of the source depth search

Figure 6.5 shows distributions of the P and T axes of the 318 focal mechanisms in Fig. 6.4. The P-axes display a consistent orientation in a northwest–southeast direction, sub-parallel to the direction of convergence between the PSP and EP. There is an obvious clockwise rotation from west to east, with the P-axes more east–west in the WF, while they are more northwest–southeast in the CR, LV, and CoR. In the southern end of the CR (region E in Fig. 6.4), there are a significant number

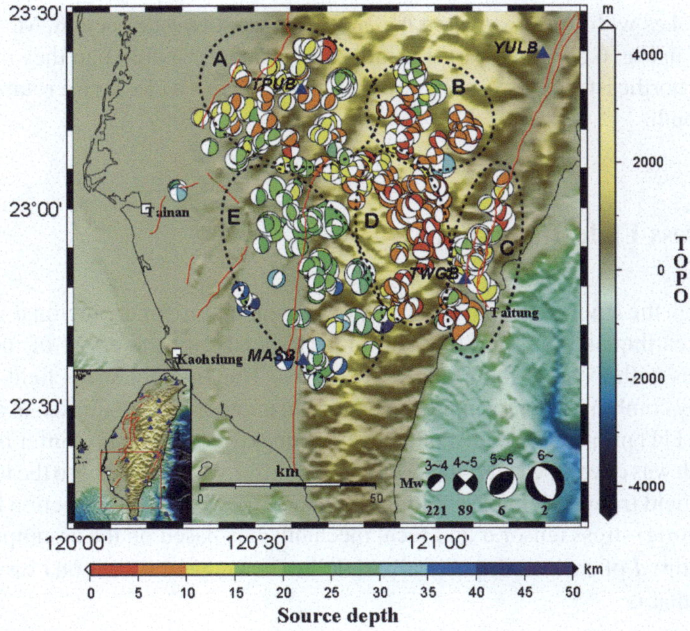

Fig. 6.4 Spatial distribution of the focal mechanisms of the 318 earthquakes used in this study. The focal mechanisms are shown by the beachballs with lower-hemisphere projection. The beachball sizes are proportional to the magnitudes, while the colors represent source centroid depth. The background color shows the topography

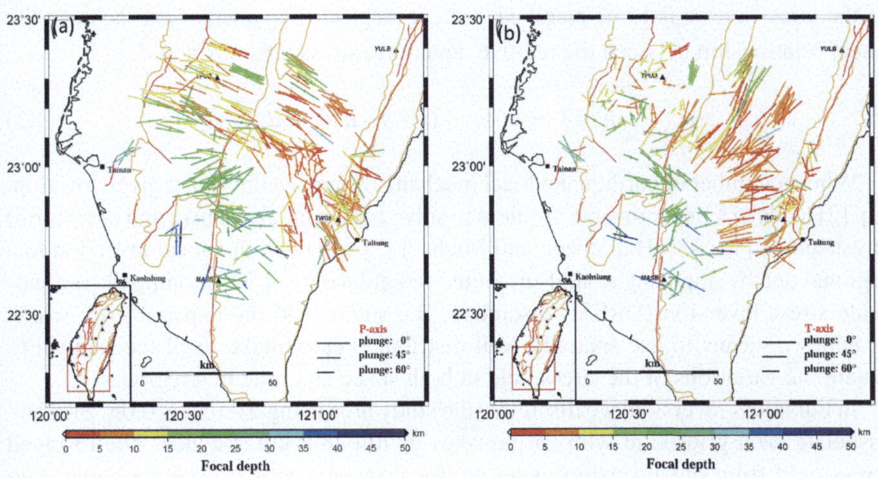

Fig. 6.5 Horizontal projections of the **a** P and **b** T axes of the focal mechanisms of the 318 earthquakes in Fig. 6.4. The directions of the line segments represent the azimuths, whereas the lengths of the line segments indicate the plunging angles of the axes. The longer the line segment, the more horizontal the axis. The center of the line segments are the epicenters of earthquakes, with the source depths shown by the colors of the line segments

of earthquakes with P-axes oriented in a northeast–southwest direction. Meanwhile, the T-axes in Fig. 6.5b show a diverse range of orientations, although they do have a consistent northeast–southwest alignment in the CR, with a clockwise rotation from north to south.

6.4 Stress Field Inversion

Even though the P and T axes in Fig. 6.5 show a clear picture of the regional variation pattern in earthquake faulting, they are more indicative of the effect of individual earthquakes with existing fault plane geometries. The ambient stress field must be obtained by combining the focal mechanisms of a number of earthquakes in a region. McKenzie [11] proposed the idea of using earthquake mechanisms to infer the stress field, which was developed later into a number of different approaches to the inversion of a stress field from fault plane solutions [1, 2, 6, 7, 12, 13]. The connection between the (deviatoric) stress tensor $\boldsymbol{\sigma}$ and focal mechanism is based on the assumption that the rake (slip) $\hat{\boldsymbol{d}}$ of each earthquake is parallel to the tangential traction $\boldsymbol{t}$ on the fault plane [1], that is:

$$\boldsymbol{t}(\hat{\boldsymbol{n}}, \boldsymbol{\sigma}) = c\hat{\boldsymbol{d}}, \tag{1}$$

where $\hat{\boldsymbol{n}}$ is the unit normal vector of the fault plane and c is a constant for all earthquakes occurring at a given location [12, 13]. When only the relative magnitude of the stress tensor is to be resolved, we can neglect the constant and achieve the linear relationship between the relative deviatoric stress tensor $\hat{\sigma}$ and $\hat{\boldsymbol{d}}$:

$$\boldsymbol{t}\left(\hat{\boldsymbol{n}}, \hat{\sigma}\right) = \hat{\sigma} \cdot \hat{\boldsymbol{n}} - (\hat{\boldsymbol{n}} \cdot \hat{\sigma} \cdot \hat{\boldsymbol{n}})\hat{\boldsymbol{n}} = \hat{\boldsymbol{d}}. \tag{2}$$

When a number of earthquake focal mechanisms are available at a given location, Eq. [2] leads to a linear inverse problem to solve for the local (relative and deviatoric) stress tensor [12, 13]. Hardebeck and Michael [7] extended this local inversion to a regional one by imposing spatial smoothness regularization. Their damped regional-scale stress inversion (DRSSI) discretizes the stress field on a spatial grid, which provides a means to use spatially well-distributed earthquake focal mechanism to obtain the variations of the stress field in both space and time in a region.

In our stress inversion, we discretize the study area using a $0.05° \times 0.05°$ grid and associate each grid point with earthquakes within an 8 km distance. The resolved stress field from our inversion using the focal mechanisms of 318 earthquakes in Fig. 6.4 is shown in Fig. 6.6.

The stress field in the WF and the CP is predominantly in a thrust environment with the maximum principle stress nearly horizontal. The azimuth is largely west–northwest, sub-parallel to the direction of plate convergence. The maximum stress also experiences a slight counter-clockwise rotation from north to south, displaying

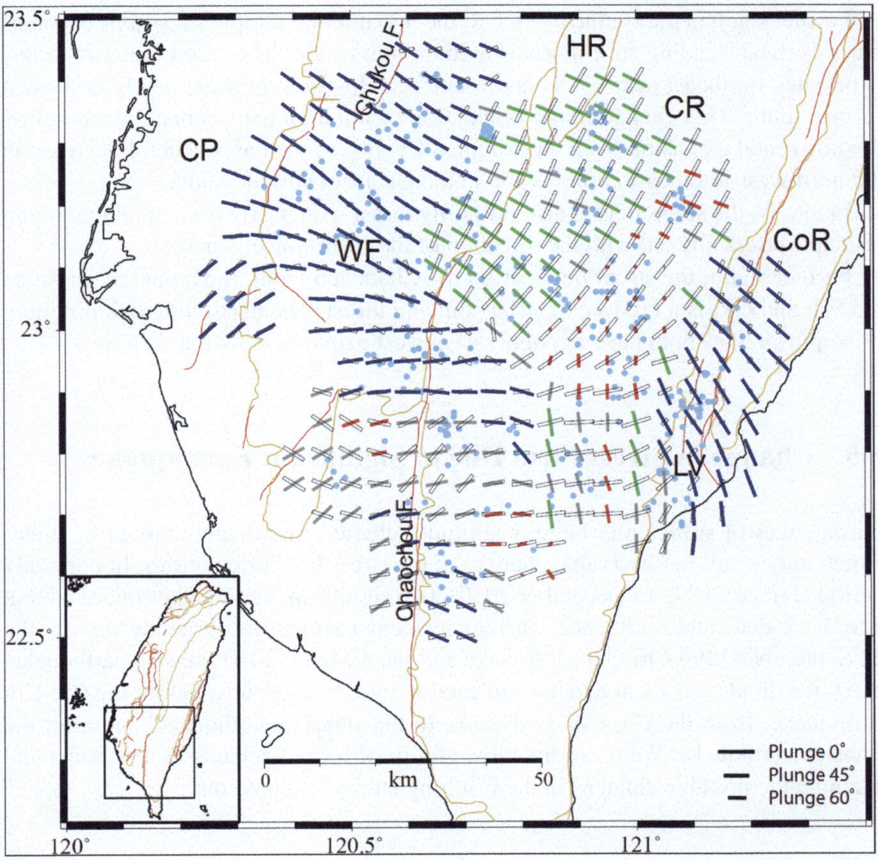

Fig. 6.6 Result of the stress inversion. This plot shows the horizontal projections of the principle stress axes, with the white line segments representing the horizontal projections of the minimum principle stress axis σ3, while the colored line segments represent the maximum principle stress axis σ1. Green, blue, red, and grey line segments represent strike-slip, thrust, normal, and mixed faulting stress environments, respectively. The directions of the line segments represent the azimuths, whereas their lengths indicate the plunging angles of the stress axes. The longer the line segment, the more horizontal the axis. The center of the line segments are the grid points used in stress inversion. The blue dots are the epicenters of the earthquakes. Abbreviations of the geological provinces are the same as in Fig. 6.1

a fan-shaped pattern as observed earlier [3–5, 8, 19]. This pattern is also similar to the one seen in the GPS velocity field (Fig. 6.1).

In the CR north of 23 °N, the minimum principle stress axis becomes more horizontal, resulting in a strike-slip stress environment. The increase in the horizontal component of the minimum stress in this part of the CR is perhaps due to the extrusion of the southern CR southward, due to the collision of the Philippine Sea Plate with Taiwan along the northern CR and LV [10].

Further south in the southeastern CR, the maximum principle stress axis becomes more vertical, leading to a normal-faulting stress field. The minimum stress axis is oriented northeast near 23 °N and rotates clockwise to become nearly east–west further south. This change in the orientation of minimum stress may be caused by the horizontal extension from an exhumed CR [8], as well as a general decrease in the northwest–southeast compression in the CR from north to south.

In the LV, the stress field is also primarily thrust with nearly horizontal maximum principle stress aligned with the direction of the plate convergence.

Further south, the stress field across the Chaochou Fault, the boundary between the WF and CR, is a mixture of strike-slip and thrust. The maximum and minimum principle stresses both have a roughly east–west azimuth, but variable plunges.

6.5 Change in Stress Field Due to Significant Earthquakes

Earthquakes of sufficiently large magnitude release a significant amount of stress, which may result in observable changes in the stress field orientations. In our study period (January 1997 to December 2010), two significant earthquakes occurred that may have detectable influence on the stress environment in our study region: the 21 September 1999 Chi-Chi earthquake and the 4 March 2010 Jiahsian earthquake. Next, we divide our focal mechanism catalog into three groups: before the Chi-Chi earthquake, from the Chi-Chi earthquake to the Jiahsian earthquake, and after the Jiahsian earthquake. We invert the three groups of focal mechanisms separately and examine the possible changes in the resulting stress field patterns.

6.5.1 Chi-Chi Earthquake

The stress inversion results using earthquakes before and after the 1999 Chi-Chi earthquake (but before the Jiahsian earthquake) are shown in Fig. 6.7. The Chi-Chi earthquake (M_W 7.7) occurred north of our study area (with an epicenter at 23.772° N, 120.982° E) and is a pure thrust on a low-angle fault plane dipping to the east–southeast direction.

The thrust stress environment in the WF and the strike-slip environment in the CR north of 23° N are largely unchanged before and after the Chi-Chi earthquake, whereas the stress patterns in the southeastern CR, LV, and across the Chaochou Fault changed dramatically after the Chi-Chi earthquake. The pattern in Fig. 6.7b clearly reveals a decrease in the horizontal component of the maximum compression in the CR after the Chi-Chi earthquake, especially around 23° N and further south. At the same time, there is an increase in the horizontal component of the maximum extension in the same region. These changes may have led to shifts in earthquake focal mechanisms from strike-slip and thrust before the Chi-Chi earthquake to normal and mixed, respectively, afterwards.

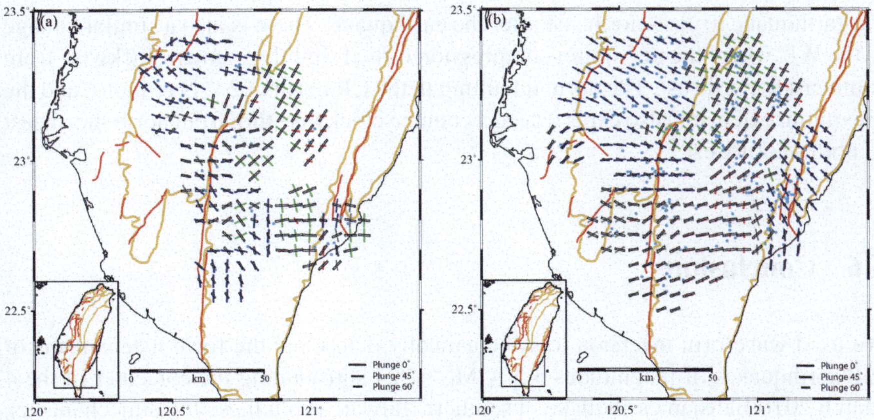

Fig. 6.7 Stress field inversion results using earthquakes **a** before and **b** after the 21 September 1999 Chi-Chi earthquake. The blue dots represent the epicenters of the earthquakes. The meanings of the line segments and colors are the same as in Fig. 6.6

6.5.2 Jiahsian Earthquake

Figure 6.8 displays the stress inversion results obtained using focal mechanisms before and after the March 2010 Jiahsian earthquake. Only eight months of data were used after the Jiahsian earthquake, with most of the earthquakes occurring across the northern portion of the Chaochou Fault. The Jiahsian earthquake (M_W 6.3) was an oblique thrust on a low-angle northeast-dipping fault plane. The maximum compression around the Jiahsian epicenter changed slightly from east to west before

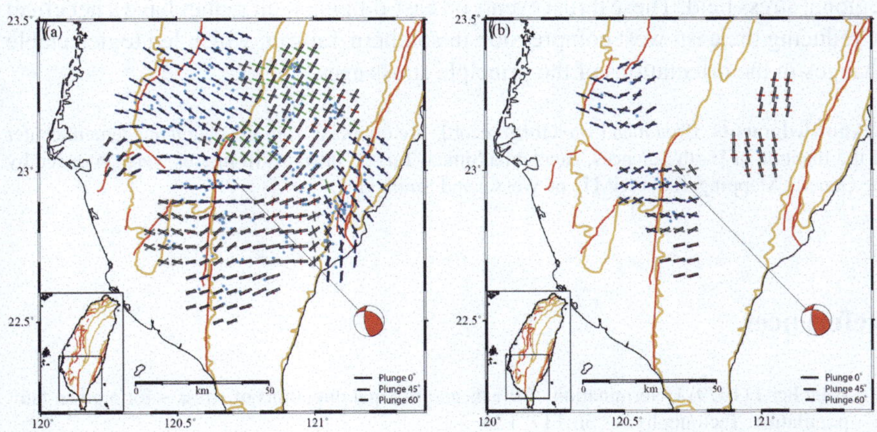

Fig. 6.8 Stress field inversion results using earthquakes **a** before and **b** after the 4 March 2010 Jiahsian earthquake. The beachball shows the focal mechanism of the Jiahsian earthquake. Other map features are the same as in Fig. 6.7

the earthquake to east–northeast after the earthquake. There is also a similar change in the WF, where the maximum compression turned slightly counter-clockwise from southeast to northeast. The normal faulting in the CR north of 23° N remains, and the maximum extension also experienced a counter-clockwise turn from north–northeast to north–northwest.

6.6 Conclusion

We used waveform inversion to systematically determine the focal mechanisms of 318 earthquakes of magnitudes $3.5 \leq M_L \leq 6.4$ surrounding the epicenter of the 4 March 2010 Jiahsian earthquake in southern Taiwan. From these focal mechanisms, we are able to obtain the orientations of the principle stress axes using a damped regional-scale stress inversion algorithm. We also carried out stress field inversion using earthquakes before and after the 1999 Chi-Chi and the 2010 Jiahsian earthquakes to further investigate the possible influence of significant earthquakes on the stress field. Our results demonstrated that there are primarily two types of stress environments in southern Taiwan: (1) the overall compression in a northwest–southeast direction in the Western Foothills and the Longitudinal Valley as a result of the convergence of the Philippine Sea and Eurasian plates, and (2) the northeast–southwest extension due to the exhumation of materials that led to the mountain building in the Central Range. These two prevailing stress conditions coupled with local differences led to areas with distinct stress characteristics: thrust in the Western Foothills and the Longitudinal Valley, strike-slip in the Central Range north of 23° N, normal in the southeastern Central Range, and mixed-type across the Chaochou Fault. The 1999 Chi-Chi and 2010 Jiahsian earthquakes had an observable influence on the regional stress field. These thrust events on east-dipping fault planes have a net effect of reducing the east–west compression in southern Taiwan, which led to detectable changes in the orientations of the principle stress axes.

Acknowledgments Broadband waveform records are obtained from the data management center of the Institute of Earth Sciences, Academia Sinica, Taipei, Taiwan. All figures were produced by the Generic Mapping Tools (GMT) of Wessel and Smith [15].

References

1. Angelier J (1979) Determination of the mean principal directions of stresses for a given fault population. Tectonophysics 56:T17–T26
2. Angelier J (1984) Tectonic analysis of fault slip data sets. J Geophys Res 89:5835–5848
3. Angelier J, Barrier E, Chu HT (1986) Plate collision and paleostress trajectories in a fold-thrust belt: the foothills of Taiwan. Tectonophysics 125:161–178
4. Bos AG, Spkman W, Nyst MCJ (2003) Surface deformation and tectonic setting of Taiwan inferred from GPS velocity field. J Geophys Res 108(B10):2458

5. Chang CP, Chang TY, Angelier J, Kao H, Lee JC, Yu SB (2003) Strain and stress field in Taiwan oblique convergent system: Constraints from GPS observations and tectonic data, Earth Planet. Sci. Lett. 214:115–127
6. Gephart JW, Forsyth DW (1984) An improved method for determining the regional stress tensor using earthquake focal mechanism data: application to the San Fernando earthquake sequence. J Geophys Res 89:9305–9320
7. Hardebeck JL, Michael AJ (2006) Damped regional-scale stress inversions: methodology and examples for southern California and the Coalinga aftershock sequence. J Geophys Res 111:B11310. https://doi.org/10.1029/2005JB004144
8. Hsu YJ, Yu SB, Simons M, Kuo LC, Chen HY (2009) Interseismic crustal deformation in the Taiwan plate boundary zone revealed by GPS observations, seismicity, and earthquake focal mechanisms. Tectonophysics 479:4–18
9. Kao H, Jian PR, Ma KF, Huang BS, Liu CC (1998) Moment-tensor inversion for offshore earthquakes east of Taiwan and their implications to regional collision. Geophys Res Lett 25:3619–3622
10. Kao H, Jian PR (2001) Seismogenic patterns in the Taiwan region: insights from source parameter inversion of BATS data. Tectonophysics 333:179–198
11. McKenzie DP (1969) The relation between fault plane solutions for earthquakes and the directions of the principal stresses. Bull Seismol Soc Am 59:591–601
12. Michael AJ (1984) Determination of stress from slip data: faults and folds. J Geophys Res 89:11517–11526
13. Michael AJ (1987) Use of focal mechanisms to determine stress: a control study. J Geophys Res 92:357–368
14. Rau RJ, Wu FT (1998) Active tectonics of Taiwan orogeny from focal mechanisms of small-to-moderate-sized earthquakes. Terr Atmos Ocean Sci 9:755–778
15. Wessel P, Smith WHF (1998) New, improved version of the generic mapping tools released. EOS Trans AGU 79:579
16. Wu YM, Chang CH, Zhao L, Shyu JBH, Chen YG, Sieh K, Avouac J-P (2007) Seismic tomography of Taiwan: improved constraints from a dense network of strong-motion stations. J Geophys Res 112:B08312. https://doi.org/10.1029/2007JB004983
17. Wu YM, Chen CC (2007) Seismic reversal pattern for the 1999 Chi-Chi, Taiwan, Mw 7.6 earthquake. Tectonophysics 429:125–132
18. Wu YM, Hsu YJ, Chang CH, Teng SY, Nakamura M (2010) Temporal and spatial variation of stress field in Taiwan from 1991 to 2007: insights from comprehensive first motion focal mechanism catalog, Earth Planet. Sci Lett 298:306–316
19. Wu YM, Zhao L, Chang CH, Hsu YJ (2008) Focal-mechanism determination in Taiwan by genetic algorithm. Bull Seismol Soc Am 98:651–661
20. Yeh YH, Barrier E, Lin CH, Angelier J (1991) Stress tensor analysis in the Taiwan area from focal mechanisms of earthquakes. Tectonophysics 200:267–280
21. Yu SB, Chen HY, Kuo LC, Lallemand SE, Tsien HH (1997) Velocity field of GPS stations in the Taiwan area. Tectonophysics 274:41–59
22. Zhao LS, Helmberger DV (1994) Source estimation from broadband regional seismograms. Bull Seismol Soc Am 84:91–104
23. Zhu L, Helmberger DV (1996) Advancement in source estimation techniques using broadband regional seismograms. Bull Seismol Soc Am 86:1634–1641
24. Zhu L, Rivera LA (2002) A note on the dynamic and static displacement from a point source in multi-layered media. Geophys J Int 148:619–627

Chapter 7
Seismo-Geological Signatures for M ≧ 7.0 Earthquakes and Their Experimental Prediction in Taiwan

Wen-Yen Chang, Hsin-Chieh Pu, and Cheng-Horng Lin

Abstract Earthquakes are one of the most serious natural disasters on Earth. A major earthquake might kill many people, ranging from several hundreds to tens of thousands in highly populated areas. Increasing population growth in high seismicity areas may transform a less significant earthquake into a major disaster. Thus, how to mitigate seismic impact is an important area of research around the world. Obviously, the seismic impact would be dramatically decreased if earthquakes could be predicted. Although earthquake prediction is still very difficult, some earthquake precursors have been reported. To study such earthquake precursors, Taiwan may be one of the best natural experimental laboratories in the world because it suffers abundant earthquakes, there is a variety of dense instrument arrays, and there are integrated programs for finding earthquake precursors. Some possible earthquake precursors were observed before the 1999 Chi-Chi earthquake (M7.3). Thus, a cocktail model is proposed here for integrating all possible earthquake precursors. The idea is to predict earthquake parameters based on cross-checking different possible precursors across the time, space, and magnitude domains. Although it might be a long time before earthquakes can be forecasted, persistent data collection and analysis are useful for preventing seismic disasters in the future.

Keywords Seismic disaster · Earthquake precursor · Dense instrument array · Taiwan

W.-Y. Chang
College of Environmental Studies, National Dong Hwa University, Hualien, Taiwan

Center for Interdisciplinary Research on Ecology and Sustainability, National Dong Hwa University, Hualien, Taiwan

H.-C. Pu
Seismological Center, Central Weather Bureau, Taipei, Taiwan

C.-H. Lin (✉)
Institute of Earth Sciences, Academia Sinica, Taipei, Taiwan
e-mail: lin@earth.sinica.edu.tw

C.-H. Lo et al. (eds.), *Earthquake Geology and Tectonophysics around Eastern Tibet and Taiwan*, Atmosphere, Earth, Ocean & Space,
https://doi.org/10.1007/978-981-15-6210-5_7

7.1 Introduction

Although the majority of geologists do not believe that earthquakes can currently be predicted, the forecasting of a large earthquake (M > 7.0) might not be impossible in the future if more consistent earthquake precursors could be clearly identified. In fact, some possible earthquake precursors (Table 7.1) have been carefully evaluated by the International Association of Seismology and Physics of the Earth's Interior (IASPEI). Among 37 nominations, five cases were selected as possible earthquake precursors [28, 29]. They include (1) foreshocks hours to months before the 4 February 1975 M7.3 Haicheng earthquake in China [26], (2) pre-shocks in the months to years before the 22 January 1988 M6.7 Tennant Creek earthquake in Australia [2], (3) seismic quiescence before strong aftershocks in Japan [19], (4) radon concentration and temperature decrease in ground water before the 14 January 1978 M7.0 Izu-Oshima-Kinkai earthquake in Japan, [24] and (5) rising ground water before the 4 August 1985 M6.1 Kettleman Hills earthquake in California, USA [20]. However, the IASPEI Sub-commission on Earth Prediction did not guarantee that those precursors could be used for earthquake prediction at the present time [29]. As a result, significant work remains before earthquakes can be forecasted, but persistent data collection and analysis are one of the most important means of identifying possible precursors.

To examine possible earthquake precursors, active seismic zones such as in China, Taiwan, Japan, California, and some other places along plate boundaries might have a better chance of yielding suitable candidates than other parts of the world owing to their frequent earthquakes. For example, some clear earthquake precursors were observed before the 1975 Haicheng earthquake (M7.3). These precursors included not only changes in ground water and soil elevations over the prior several months, but also unusual animal behavior and some foreshocks [9, 25]. Following this earthquake, researchers in China and many other nations have been continuously looking for a variety of possible earthquake precursors. Unfortunately, there are still no precursors that can be consistently detected until now.

Since Taiwan is located along the plate boundary between the Eurasian Plate and the Philippine Sea Plate, many earthquakes have taken place within this small

Table 7.1 The preliminary list of significant earthquake precursors selected by IASPEI in 1994

	Reference	Precursor	Earthquake
1	Wu et al. [26]	Foreshocks	Haicheng earthquake (M7.3) in China, 1975
2	Bowman [2]	Pre-shocks	Tennant Creek earthquake (M6.7) in Australia, 1988
3	Matsuura [19]	Seismic quiescence	Several earthquakes in Japan
4	Wakita et al. [24]	Radon and temperature decrease	Izu-Oshima-Kinka earthquake (M7.0) in Japan, 1978
5	Roeloffs and Quilty [20]	Rising ground water	Kettleman Hills earthquake (M6.1) in California, 1985

area. Among them, some large earthquakes (M > 7.0) have occasionally occurred in the past and will inevitably occur again in the future. For instance, the 1999 Chi-Chi earthquake (M7.3) killed more than 2400 people in Taiwan. For mitigating any seismic impact, several dense instrument arrays such as seismic networks, GPS arrays, and many other physical observation sites have been installed in Taiwan since 1990. As a result, there is an improved opportunity to search for possible earthquake precursors through a variety of observations and analyses. Some integrated research projects have been conducted in Taiwan and a variety of interesting earthquake precursors have been reported in past decades.

7.2 General Tectonics and Seismicity

The island of Taiwan is located at a small portion of the convergence zone between the Eurasian Plate (EUP) and Philippine Sea Plate (PSP). The plate interaction is clearly demonstrated by the abundant earthquakes in the Taiwan area (Fig. 7.1). The PSP is subducting beneath the EUP along the Ryukyu trench in northeastern Taiwan, while the EUP in under-thrusting the PSP along the Manila Trench in southern Taiwan. Two subduction slabs are unambiguously marked by the clustering earthquakes within the subduction zone in northeastern and southern Taiwan, respectively. A significant collision orogeny has been taking place in Taiwan due to the strong convergence (7–8 cm/year) between the two plates since the late Miocene [21, 31]. The suture zone between the two plates is marked along the Longitudinal Valley. East of the suture zone, the Coastal Range is part of the Luzon Volcanic Arc in the PSP. West of the suture, the major geological units from west to east, including the Coastal Plain, Western Foothills, Hsuehshan Range, Backbone Ridge, and Eastern Central Range, belong to the continental margin of the EUP. Both the topographic relief and metamorphic rate of the geological units in the EUP gradually increase eastward. A major mountain-building process is taking place across most of the geological units, and a significant topographic relief (up to 3940 m) is created as a result of the strong horizontal shortening and vertical exhumation.

In addition to abundant earthquakes, strong crustal deformation is shown by a number of active faults on the surface (Fig. 7.2). In total, 31 active faults have been identified by the Central Geological Survey of Taiwan [12]. Most of the active faults are distributed in the western part of Taiwan (Western Plain or Western Foothills) and the Longitudinal Valley (the suture zone between the two plates). The general distribution of those active faults is essentially consistent with the locations of disastrous earthquakes in the past century [8] (Fig. 7.2). However, a detailed investigation of some recent earthquakes shows the disastrous earthquakes might be associated with the identified faults, as well as some blind faults in the Taiwan area.

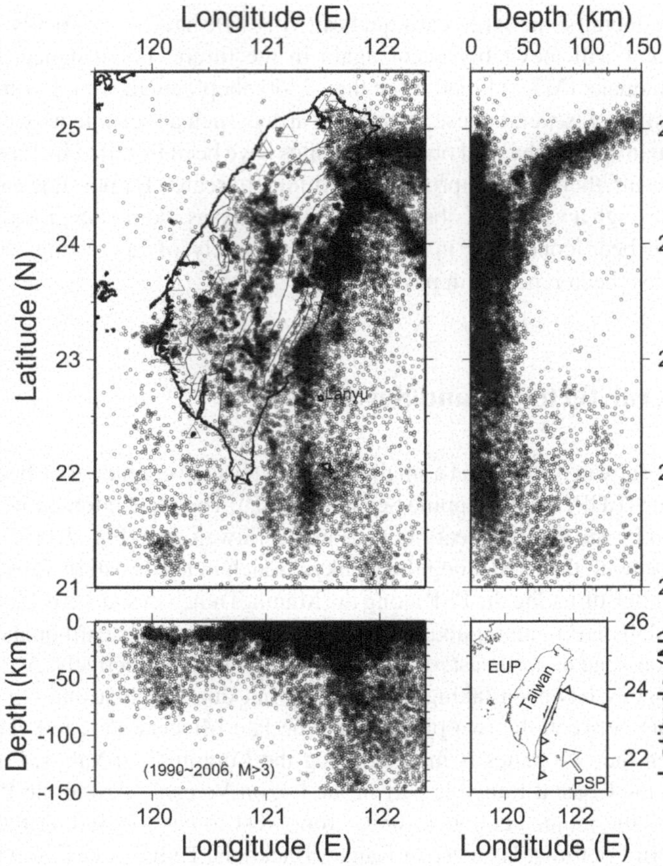

Fig. 7.1 Background seismicity (circles) and the seismic stations (triangles) of the central weather bureau seismic network in the Taiwan area

7.3 Study of Earthquake Precursors in Taiwan

To study possible earthquake precursors, Taiwan might be one of the best natural experimental laboratories in the world based on the following considerations. First, the frequent large earthquakes occurring in Taiwan provide many opportunities to check possible earthquake precursors. The seismicity in the Taiwan area is extremely high, owing to the strong plate convergence between the Eurasian and Philippine Sea Plates (Fig. 7.1). On average, earthquakes with a magnitude greater than five ($M > 5$) occur more than ten times per year (Fig. 7.3). Among them, two or three earthquakes may have magnitudes greater than 6. In other words, almost every month, we can expect a strong earthquake ($M > 5$) in the Taiwan area (E119.5–122.5°; N21–25.5°). Thus, possible earthquake precursors from those large earthquakes can be regularly examined within a short period in Taiwan.

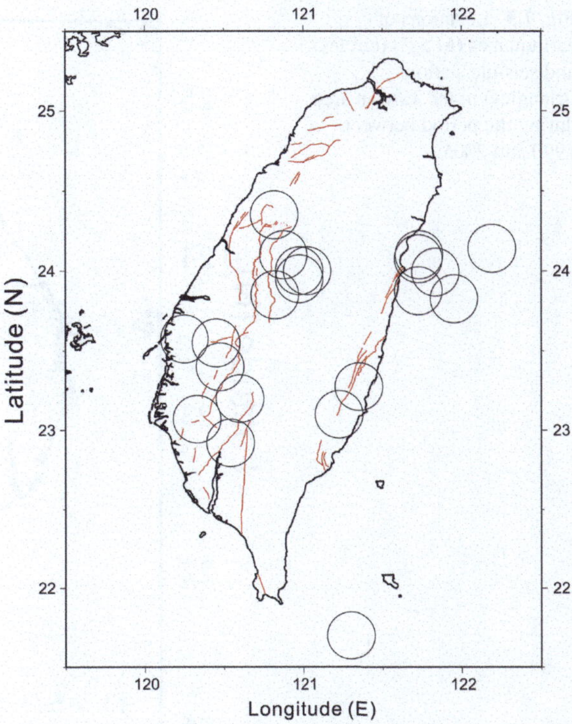

Fig. 7.2 Locations of the earthquake sequences that caused the ten largest disasters before 1998 [8] and four major earthquakes after 1999 in the Taiwan area. The active faults are marked with red lines

Second, many dense instrument arrays, including seismometers (Fig. 7.4), GPS (Fig. 7.5), strain-meters, ground water wells, and so on, have been deployed in the Taiwan area for examining possible precursors. As we do not know where an earthquake will occur, these dense instrument arrays provide a strong opportunity to search for possible precursors nearby an epicenter. Third, many well-trained scientists and experts, including seismologists, geologists, and geo-physicists have focused on the geoscientific issues in the Taiwan area. Many powerful methods and techniques have been developed during the past decades. Thus, significant quantities of valuable data and analyzed results have been carefully collected and obtained. Among them, some earthquake precursors have been carefully examined in the past decades.

Third, several integrated research projects have been conducted in the Taiwan area in the past decades. The first integrated earthquake prediction research program in Taiwan started in 1979 [23]. This program included the study of (1) spatial and temporal seismicity variations, (2) trilateration surveys, (3) gravity surveys, (4) geomagnetic anomalies, and (5) changes in radon content at hot springs. Some interesting results were observed and led to further studies. After the Chi-Chi earthquake (M7.3), a sequence of research programs entitled "the integrated Search for Taiwan Earthquake Precursors" (iSTEP) started in 2002 [16]. In addition to possible traditional precursors such as seismological variations, potential fields, and surface deformation, a new study of ionospheric total electronic content (TEC) anomalies

Fig. 7.3 Locations of earthquakes (M > 5) (circles) and seismic stations (triangles) in the Taiwan area during the period between 1990 and 2005

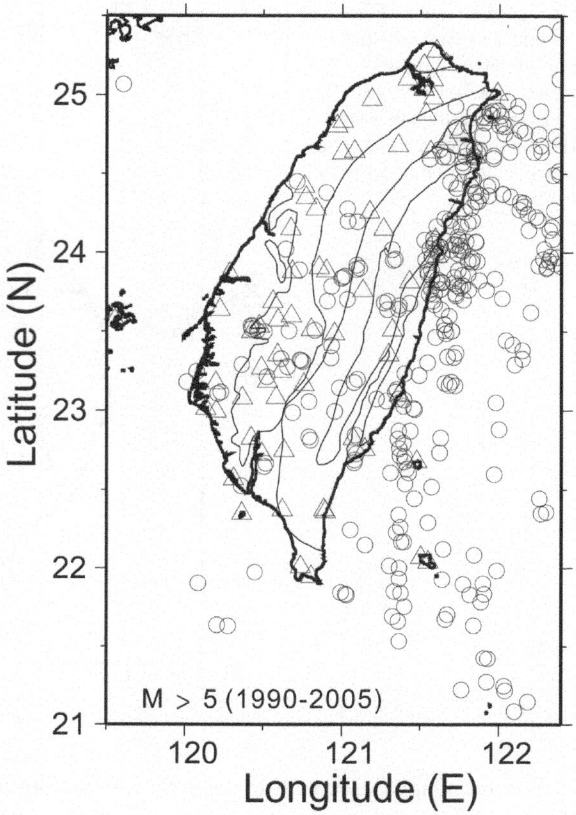

was developed. The statistical analyses showed that TEC anomalies were most likely to be a credible precursor [16]. Meanwhile, another integrated earthquake precursor program began in 2013, titled "Integrated Earthquake Precursors and Early Warning for Seismic Disaster Prevention in Taiwan". This program, mainly supported by the National Science Council in Taiwan, integrated a variety of studies on earthquake precursors such as gravity and geomagnetic surveys, telluric electromagnetics, down-hole strain-meters, seismicity, ground water monitoring, geochemical analyses, GPS surveys, ionospheric density, thermal infrared rays, gamma radiation detection, and theoretical modeling (Fig. 7.6). Some valuable results have been observed [4, 7,10, 9–15, 27, 5].

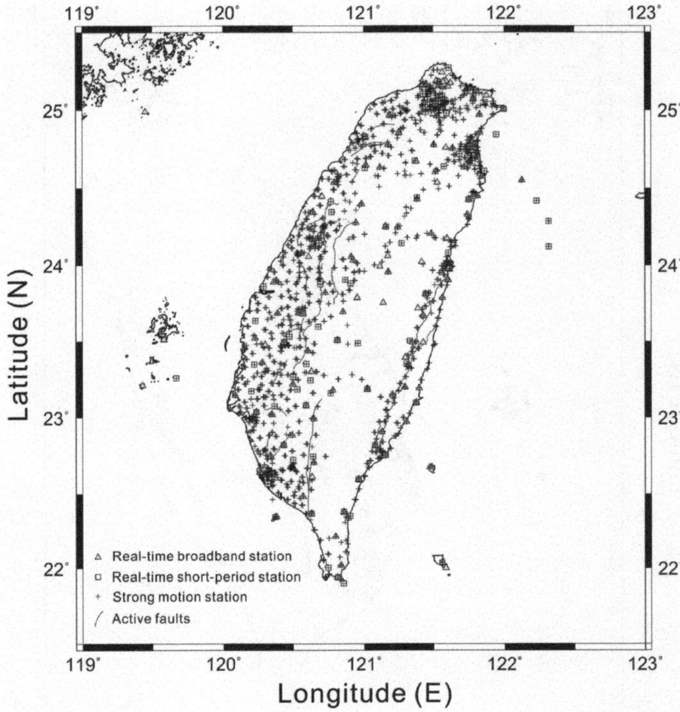

Fig. 7.4 Various seismic stations including broadband (triangle), short-period (square), and strong motion (plus sign) in the Taiwan area

7.4 Possible Precursors Before the Chi-Chi Earthquake

The 1999 Chi-Chi earthquake (M7.3) was not only one of the largest earthquakes in the Taiwan area in the past century, but also was well recorded by dense instrument arrays, owing to its epicenter being located almost in the center of Taiwan. After detailed examinations of different kinds of seismic, geodetic, geophysical, geochemical, and geological data in Taiwan, some possible earthquake precursors have been claimed. The following is a brief summary of some possible precursors before the Chi-Chi earthquake.

First, the spatial and temporal variations of earthquake activity were unambiguously detected before the Chi-Chi earthquake [7, 5, 27]. The accelerating seismicity of moderate-sized earthquakes was observed before the 1999 Chi-Chi earthquake. Furthermore, a seismic quiescence was detected before the Chi-Chi earthquake. Those temporal variations were also detected by seismic b-values, which showed a significant decrease several months before the Chi-Chi earthquake.

Second, temporal variations of the ionospheric total electron content (TEC) were detected several days before the Chi-Chi earthquake [11, 14]. The TEC derived from dense GPS sites in the Taiwan area provides reliable information for examining

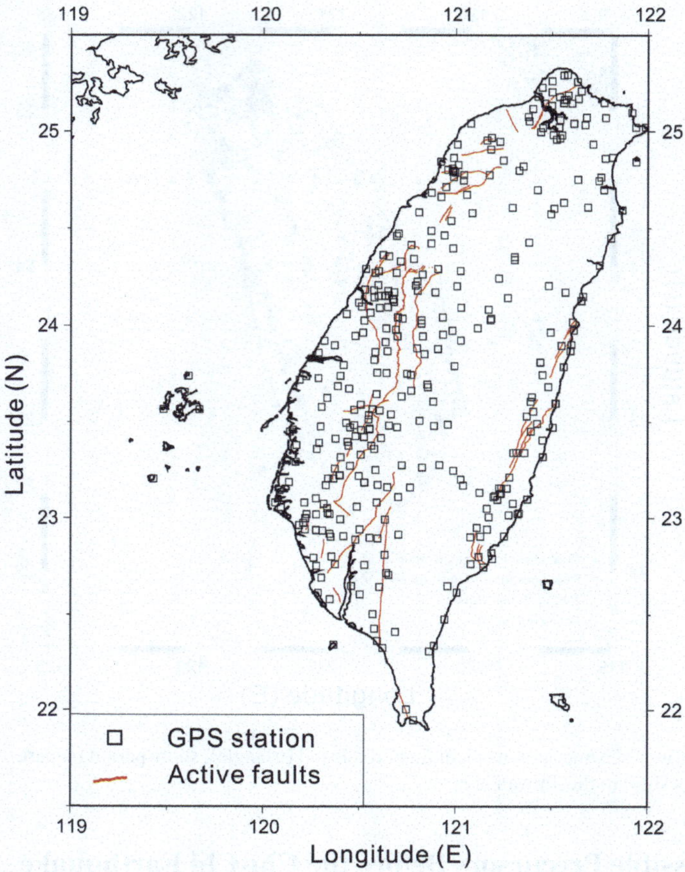

Fig. 7.5 GPS stations and surface traces of the active faults in the Taiwan area

possible anomalies before large earthquakes. In fact, they have been claimed as possible earthquake precursors for many large earthquakes in Taiwan [15], China [13], Japan [18], and many other places.

Third, unusual decreases of water levels were consistently observed in most wells in the footwall side of the Chelungpu Fault in central Taiwan several months before the Chi-Chi earthquake [4]. In addition, a coupling of pre-earthquake anomalous phenomena between long-term ground water levels and surface strain derived from GPS stations was identified before the Chi-Chi earthquake in central Taiwan.

Fourth, geomagnetic data recorded at one station (Liyutan), approximately 8 km from the rupture fault, showed significant fluctuations for more than a month prior to the earthquake [30]. In addition, the ULF magnetic emission data recorded at Lunping, which is 120 km from the epicenter, showed a significant enhancement for two months before the Chi-Chi earthquake [1]. In addition to geomagnetic anomalies

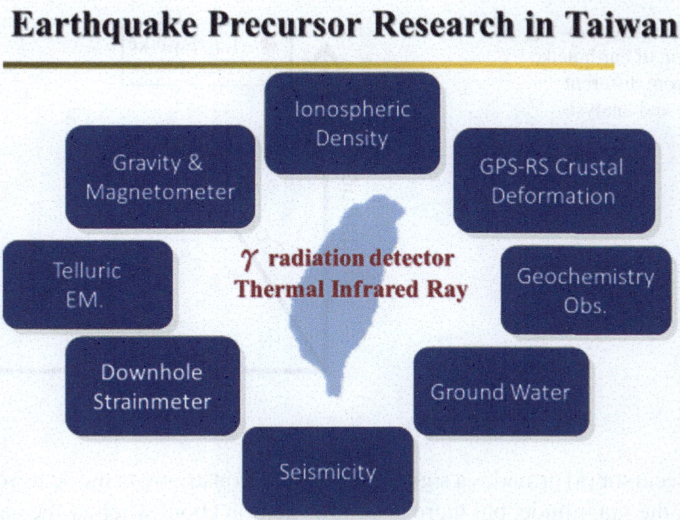

Fig. 7.6 The integration of earthquake precursor studies in the Taiwan

before the Chi-Chi earthquake, potential relationships between seismo-conductivity and seismo-deformation were reported during two M6 earthquakes in Taiwan [3, 6].

Fifth, possible short-term precursors (with a length of several days) of slow-slip events were detected before the Chi-Chi earthquake [10]. These slow-slip events derived from velocity seismograms recorded at broadband seismic stations near the Chi-Chi earthquake probably occurred along the fault plane in the upper crust.

Finally, an integration of possible earthquake precursors, including seismological variations, geomagnetic field changes, ground surface deformation, and ionospheric variations, associated with the Chi-Chi earthquake and its aftershocks has been summarized under the iSTEP program in Taiwan [22].

7.5 Integration of Earthquake Precursors

Integration of several possible precursors might be one of the best strategies to achieve the goal of earthquake prediction. A successful earthquake prediction needs to give the earthquake's location, time, and magnitude simultaneously. Unfortunately, it is still very difficult to identify these parameters from any single observation or analyzed result. In fact, based on the previous results mentioned above, some earthquake precursors might be more sensitive at the earthquake location while others are relative to either the occurrence time or magnitude of the earthquake. Therefore, a cocktail model is proposed for integrating all possible earthquake precursors (Fig. 7.7). The general idea is to predict the earthquake parameters based on cross-checking different possible precursors across the time, space, and magnitude domains. For instance, in

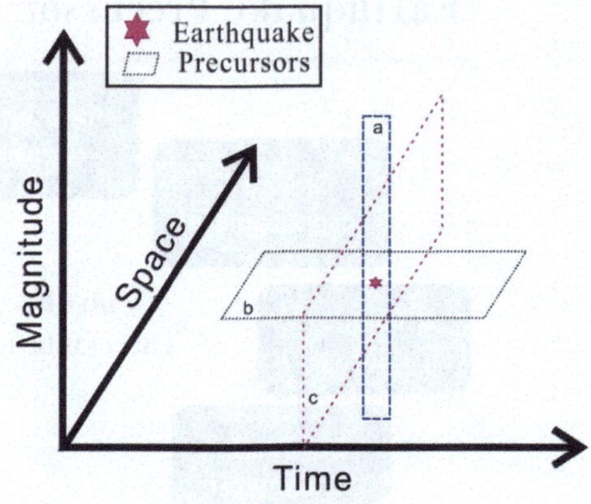

Fig. 7.7 Schematic showing the integration of earthquake precursors from different observations and analysis results

Fig. 7.7, precursor (a) provides a significantly better constraint on the occurrence time rather than the magnitude, but there is no information about whether the earthquake occurs. Precursor (b) has a better constraint on the location than the occurrence time, yet information on the earthquake magnitude is lacking. Precursor (c) in Fig. 7.7 provides some information on both earthquake location and magnitude, but does not yield any information on when the earthquake will occur. If we combine the three different precursors (a, b, and c) together, then the earthquake might be predictable. In other words, by using multiple observations of earthquake anomalies, we may be able to estimate when and where the next large earthquake may occur. By evaluating which anomalies are significant and determining their weightings, we may be able to develop an earthquake prediction analysis.

To collect different possible earthquake precursors at one site, the E-TEC (Eastern Taiwan Earthquake Research Center) was established and officially inaugurated in 2013 at National Dong Hwa University in Hualien County. The center aims to provide an integrated platform for researchers and scientists to conduct new research on earthquake precursors and early warnings for seismic disaster prevention in eastern Taiwan. This center has multiple functions and important roles. Eastern Taiwan is a good location for an earthquake research facility as temblors occur frequently in the East Taiwan rift valley. The center can collect intensive real-time data and strengthen monitoring mechanisms and research on the region's seismic activity. We intend to integrate multi-disciplinary observations, with the E-TEC equipped with stations to monitor a wide array of factors used by seismologists to detect possible earthquake precursors, including seismicity, GPS, strain-meters, ground water, geochemistry, gravity, electromagnetic and ionospheric density, infrared remote sensing, gamma radiation, and thermal infrared, and so on. E-TEC will maximize the amount and value of data for researchers and we hope that the range of monitoring equipment will enable them to predict where and when the next big earthquake will strike Taiwan, as well as determine which precursors are the most accurate for developing reliable earthquake prediction models.

7.6 Summary

Major earthquakes have caused many disasters around the world, and may become more significant as populations grow in high seismicity areas. Although earthquake prediction might still be a long way off, the study of earthquake precursors through persistent data collection and analysis might help in the future mitigation of seismic impact. Taiwan might be one of the best natural experimental laboratories in the world for investigating possible earthquake precursors as Taiwan has abundant earthquakes, many dense instrument arrays, and some long-term integrated programs for finding earthquake precursors. A careful review of all relevant data in the Taiwan area shows some possible earthquake precursors were reported before the 1999 Chi-Chi earthquake (M7.3). The possible precursors included spatial and temporal seismicity, ionospheric electron content, water levels, geomagnetic observations, and slow-slip events. Thus, a cocktail model was proposed for integrating all possible earthquake precursors. The goal is to predict earthquake parameters based on cross-checking different possible precursors across the time, space, and magnitude domains.

References

1. Akinaga Y, Hayakawa M, Liu JY, Yumoto K, Hattori K (2001) A precursory ULF signature for the Chi-Chi earthquakes in Taiwan. Nat Hazard Earth Syst Sci 1:33–36
2. Bowman JR (1997) Case 22: A seismicity precursor to a sequence of M = 6.3–6.7 midplate earthquakes in Australia. Pure Apply Geophys 149:61–78
3. Chen CH, Tang CC, Cheng KC, Wang CH, Wen S, Lin CH, Wen YY, Meng G, Yeh TK, Jan JC, Yen HY, Liu JY (2015) Groundwater-strain coupling before the 1999 Mw 7.6 Taiwan Chi-Chi earthquake. J Hydrol 524:378–384
4. Chen C-H, Wang CH, Wen S, Yeh TY, Lin CH, Liu JY, Yen HY, Lin C, Rau RJ, Lin TW (2013) Anomalous frequency characteristics of groundwater level before major earthquakes in Taiwan. Hydrol Earth Syst Sci 17:1693–1703
5. Chen CC (2003) Accelerating seismicity of moderate-size earthquakes before the 1999 Chi-Chi earthquake: testing time-prediction of the seld-organizing spinodal model of earthquakes. Geophys J Int 155:F1-F5
6. Chen CH, Lin CH, Wang CH, Liu JY, Yeh TK, Yen HY, Lin TW (2015) Potential relationships between seismo-deformation and siesmo-conductivity anomalies. J Asian Earth Sci 114, 327–337
7. Chen CC, Rundle JB, Holliday JR, Nanjo KZ, Turcotte DL, Li SC, Tiampo KF (2005) Geophys Res Let 32:L22315
8. Cheng SN, Yeh YT, Hsu MT, Shin TC (1999) Photo album of ten disastrous earthquakes in Taiwan, publishedjointlybytheCentral Weather Bureau and the Institute of Earth Sciences. Academia Sinica, Taipei, Taiwan, pp 290
9. Jones LM, Wang B, Zu S, Fitch TJ (1982) The foreshock sequence of the February 4, 1975. Haicheng earthquake (M = 7.3), J.G.R. 87:4575–4584
10. Lin CH (2012) The possible observation of slow slip events prior to the occurrence of the 1999 Chi-Chi earthquake. Terr Atm Ocean 23:145–159
11. Lin JW (2013) Taiwan' Chi-Chi earthquake precursor detection using nonlinear principal component analysis to multi-channel total electron content records. J Earth Sci 24:244–253

12. Lin C-W, Chang H-C, Lu S-T, Shih T-S, HuangW-J (2000) An introduction to the active faults of taiwan, 2nd ed., Explanatory text of the active fault map of Taiwan (in Chinese with English abstract). Spec Pub Cent Geol Surv 13, pp 122, Taipei, Taiwan
13. Lin CH (2009) Foreshock characteristics in Taiwan: potential earthquake warning, J Asian Earth Sci 34(5):655–662
14. Liu JY, Chen YI, Chuo YJ, Tsai HF (2001) Variations of ionospheric total electron content during the Chi-Chi earthquake. Geophys Res Lett 28:1383–1386
15. Liu JY, Chen YI, Pulinets SA, Tsai YB, Chou YJ (2000) Seismo-ionospheric signatures prior to M > 6.0 Taiwan earthquakes. Geophys Res Lett 27:3113–3116
16. Liu JY, Tsai YB, Chen CH, Chen YI, Yen HY (2016) Integrated search for Taiwan earthquake precursors (ISTEP). IEEJ Trans Fund Mater 136(5):214–220
17. Liu JY et al (2009) Seismo-ionospheric GPS total electron content anomalies observed before the 12 May 2008 Mw7.9 Wenchuan earthquake. J.G.R. 114:A04320
18. Liu JY et al Ionospheric disturbances triggered by the 11 March 2011 M9.0 Tohoku earthquake, J.G.R. 116:A06319
19. Matsuura RS (1986) Precursor quiescence and recovery of aftershock activities before some large aftershocks. Bull Earthq Res Inst Univ Tokyo 61:1–65
20. Roeloffs E, Qualty E (1997) Case 21: water level and strain changes preceding and following the August 4, 1985 Kettleman Hills, California, earthquake. Pure Apply Geophys 149:21–60
21. Seno T (1977) The instantaneous rotation vector of the Philippine Sea Plate relative to the Eurasian Plate. Tectonophysics 42:209–226
22. Tsai YB, Liu JY, Ma KF, Yen HY, Chen KS, Chen YI, Lee CP (2006) Precursory phenomena associated with the 1999 Chi-Chi earthquake in Taiwan as identified under the iSTEP program. Phys Chem Earth 31:365–377
23. Tsai YB, Teng TL, Yeh YH, Yu SB, Liu KK, Wang JH (1983) status of earthquake prediction research in Taiwan, ROC. Bull Inst Earth Sci 3:1–26
24. Wakita H et al (1980) Radon anomaly: a possible precursor of the 1978 Izu-Oshima-Kinkai earthquake. Science 207:882–883
25. Wang K, Chen QF, Sun S, Wang A (2006) Predicting the 1975 Haicheng earthquake. BSSA 96:757–795
26. Wu KT et al (1978) Certain characteristics of the Haicheng earthquake (M = 7.3) sequence. Chin Geophys 1:289–308
27. Wu YM, Chiao LY (2006) Seismic quiescence before the 1999 Chi-Chi earthquake, Taiwan, Mw7.6 earthquake. BSSA 96:321–327
28. Wyss M (1991) Evaluation of proposed earthquake precursors. AGU, Washington, DC, p 94
29. Wyss M (1997) Second round of evaluations of proposed earthquake precursors. Pure Appl Geophys 149:3–16
30. Yen HY, Chen CH, Yeh YH, Liu JY, Lin CR, Tsai YB (2004) Geomagnetic fluctuations during the 1999 Chi-Chi earthquake in Taiwan. Earth Planets and Space 56:39–45
31. Yu SB, Chen HY, Kuo LC (1997) Velocity field of GPS stations in the Taiwan area. Tectonophysics 274:41–5

Chapter 8
Geophysical Evidence for the Ancient Suture Within the Cathaysia Block

Hui Teng Cai, Hao Kuo-Chen, Wen-Bin Doo, Chung-Pai Chang, Xin Jin, Yao-Wen Kuo, Chien-Ying Wang, Wen-Shan Chen, and Wen-Yen Chang

Abstract The Zhenghe-Dapu fault zone, which has a length of ~800 km and is located in southeastern China, has been mapped as a suture. Geological and geochronological data indicate that the suture is filled with chaotic pelagic sedimentary rocks and ophiolite mélange that separated the Cathaysia block at ~860 Ma by rifting and later closed as one block at ~240 Ma. However, the crustal structure corresponding to this ancient suture has seldom been reported. In this study, we investigate this suture based on four east–west reflected/refracted P-wave velocity structures, along with data of Bouguer and magnetic anomalies in Fujian province in China. The results indicate that the suture has imprinted in the crustal structure, as shown by the low velocity zone and thicker crust beneath the suture as well as the low Bouguer anomalies. Additionally, a negative magnetic anomaly zone is located beneath the Zhenghe-Dapu fault zone and is consistent with a low-Vp velocity zone. This fault zone is suggested to be a rift basin filled with chaotic sedimentation. The

H. T. Cai · X. Jin
College of Civil Engineering, Fuzhou University, Fuzhou, China

Earthquake Administration of Fujian Province, Fuzhou, China

H. Kuo-Chen (✉) · C.-P. Chang · Y.-W. Kuo · C.-Y. Wang
Department of Earth Sciences, National Central University, Zhongli, Taiwan
e-mail: kuochen@g.ncu.edu.tw

W.-B. Doo
Center for Environmental Studies, National Central University, Zhongli, Taiwan

C.-P. Chang
Center for Space and Remote Sensing Research, National Central University, Zhongli, Taiwan

W.-S. Chen
Department of Geosciences, National Taiwan University, Taipei, Taiwan

W.-Y. Chang
Department of Natural Resources and Environmental Studies, National Dong Hwa University, Hualien, Taiwan

Center for Interdisciplinary Research on Ecology and Sustainability, National Dong Hwa University, Hualien, Taiwan

C.-H. Lo et al. (eds.), *Earthquake Geology and Tectonophysics around Eastern Tibet and Taiwan*, Atmosphere, Earth, Ocean & Space,
https://doi.org/10.1007/978-981-15-6210-5_8

compressional force may therefore have caused these low-susceptibility and low-density materials to move to deeper depths in Fujian province. These geological and geophysical data indicate that this suture is an important feature in southeastern China and they advance our knowledge of plate reconstruction.

Keywords Seismic reflected/refracted profile · Gravity anomaly · Magnetic anomaly · Zhenghe-Dapu fault zone · Suture zone · Cathaysia block

8.1 Introduction

The crustal evolution of individual domains in the Precambrian era is of key importance in geodynamic reconstructions. The South China Continent (SCC) has a complex tectonic history, and consists of two major Precambrian continental blocks: the Yangtze block to the northwest and the Cathaysia block to the southeast, which is defined by the suture of the Shaoxing-Jiangshan-Pingxiang fault, with a length of ~1200 km (Fig. 8.1); [3, 16]. In contrast, within the Cathaysia block, the Zhenghe-Dapu fault with a length of ~800 km is a suture that separates the block into two micro-continents (Fig. 8.1) [3, 17]. However, due to the scale of the fault, along with geological and geochemical evidence, the behavior of the Shaoxing-Jiangshan-Pingxiang fault is better documented than that of the Zhenghe-Dapu fault [17]. The Zhenghe-Dapu fault has been recognized to be a suture in some geological and geochemical research papers [3, 5, 13, 14, 17]. However, few papers have reported geophysical data.

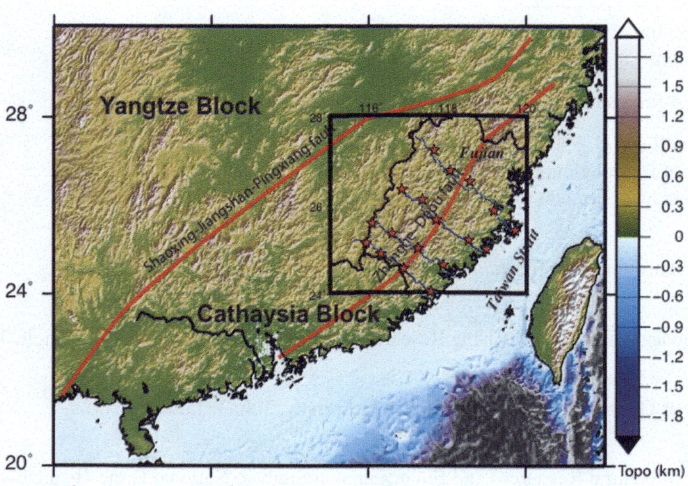

Fig. 8.1 Tectonic setting in southeastern China. Rectangular box: study region; red stars: explosions; blue triangles: seismic stations; and red lines: sutures

In the last decade, however, geophysical data sets have grown considerably and the resolution of subsurface imaging has improved, thanks to decreases in the cost of instruments and in computation time. For instance, the suture zone (Shaoxing-Jiangshan-Pingxiang fault, Fig. 8.1) is shown by the offset of the Moho depth and velocity contrast between the Yangtze and Cathaysia blocks [6, 9, 21]. A study on receiver function shows that the crustal thickness is larger in the Yangtze block (approximately 40–56 km) than in the Cathaysia block (approximately 25–34 km) [6]. Legendre et al. [9] also revealed using teleseismic surface tomography that the velocity structure of the Yangtze block is different from that of the Cathaysia block. Moreover, based on 2D seismic and density crustal profiles, Zhao et al. [21] showed that the rock properties of these two blocks are very different. Some geophysical evidence has also been reported for the Zhenghe-Dapu suture within the Cathaysia block: two independent studies of travel-time tomography [2, 8] indicated an anomalous low velocity beneath the Zhenghe-Dapu fault zone, and He et al. [6] also showed a similar Moho topography (the Moho is thicker beneath the fault zone and shallower toward the east and west). In this study, to better describe the current crustal structure of this closed rift basin or the suture zone, we resolve seismic reflected/refracted profiles from active sources and reveal the gravity and magnetic anomalies beneath this suture in Fujian province.

This area in Fujian province has developed a series of faults and a series of undulate hills. There are three major NNE-oriented faults: the Changle-Zhaoan fault zone, the Zhenghe-Dapu fault zone, and the Shaowu-Heyuan fault zone. Geological and chronological studies have shown that the Zhenghe-Dapu fault zone is an ancient suture zone that separates the Early Paleozoic fold belt to the west from the Late Mesozoic magmatic belt to the east on the surface [13]. The tectonic/geological units can be defined in three large units from east to west as follows (Fig. 8.2): (1) the Jurassic–Cretaceous magmatic arc of the coastal southeast, composed of volcanic arc, basin, and magmatic complex rocks (FJ-E); (2) the suture zone of central Fujian, composed of the Paleozoic pelagic sedimentary rocks, basaltic debris, and ophiolite mélange (FJ-C); and (3) the Cathaysia massif, the Paleozoic inter- and back-arc basin, and rift basin (FJ-W) [13].

8.2 Seismic Reflected/Refracted Profiles

The Across Taiwan Strait Explosion Experiment (ATSEE) was an onshore–offshore seismic experiment conducted from 2010 to 2012 by the Fujian Earthquake Administration of Fujian province and in collaboration with Taiwanese scientists to obtain the crustal structure in the Taiwan Strait and Fujian province for the purpose of seismic hazard prevention. During the experiment, there were a total of eighteen explosions with eight linear arrays (four EW and four NS arrays) of receivers deployed (Fig. 8.2). Because of the dense spacing (~ 2 km) of the seismic arrays, the refracted and reflected phases (Pg, Pn, PcP, and PmP) are clearly identified. Using the refracted arrivals Pg and Pn, we carried out a 2D tomographic inversion using Hole's code [7]. This is

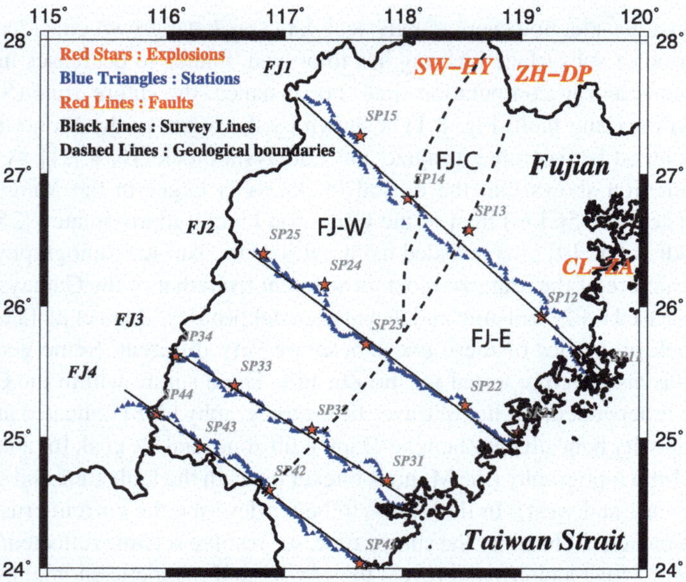

Fig. 8.2 Geological boundaries, fault zones, and the geometry of the active source experiment. Dashed lines: geological boundaries; red lines: fault zones; red stars: explosions; blue triangles: seismic stations; black lines: seismic survey lines; SW-HY: the Shaowu-Heyuan fault zone; CL-ZA: the Changle-Zhaoan fault zone; ZH-DP: the Zhenghe-Dapu fault zone; FJ-E, FJ-C, and FJ-W: see text for a detailed description

preliminary result is the initial model input for the later forward modeling. Once the preliminary velocity distribution is built, we performed 2D forward modeling using a ray-tracing method with the reflected and refracted phases [18, 19]. The forward modeling starts by fitting the observed refracted arrival times. We then model the reflections, by focusing on locating the depths of the reflectors. Forward modeling was performed from the top to the bottom reflectors, resulting in an optimal final model.

As a result, the residuals of the final models among the four transects are below 0.2 s (Figs. 8.3, 8.4, 8.5 and 8.6). The mid-crustal reflectors are present in all of the transects at a depth of ~20 km. It is interesting to note that the depth of the mid-crustal reflection layer deepens toward the center of Fujian beneath the Zhenghe-Dapu fault, especially in transects FJ2, FJ3, and FJ4. The Moho depth is approximately 30–35 km and has a similar pattern to that of the mid-crustal reflection (i.e., it becomes deeper beneath the Zhenghe-Dapu fault). Furthermore, He et al. [6] and Cai et al. [2] showed a similar Moho topography based on receiver function and travel-time tomography methods, respectively. However, in this study, we resolve the crustal structure in much more detail based on better mid-crustal and Moho reflection constraints from the reflected phases (PcP and PmP).

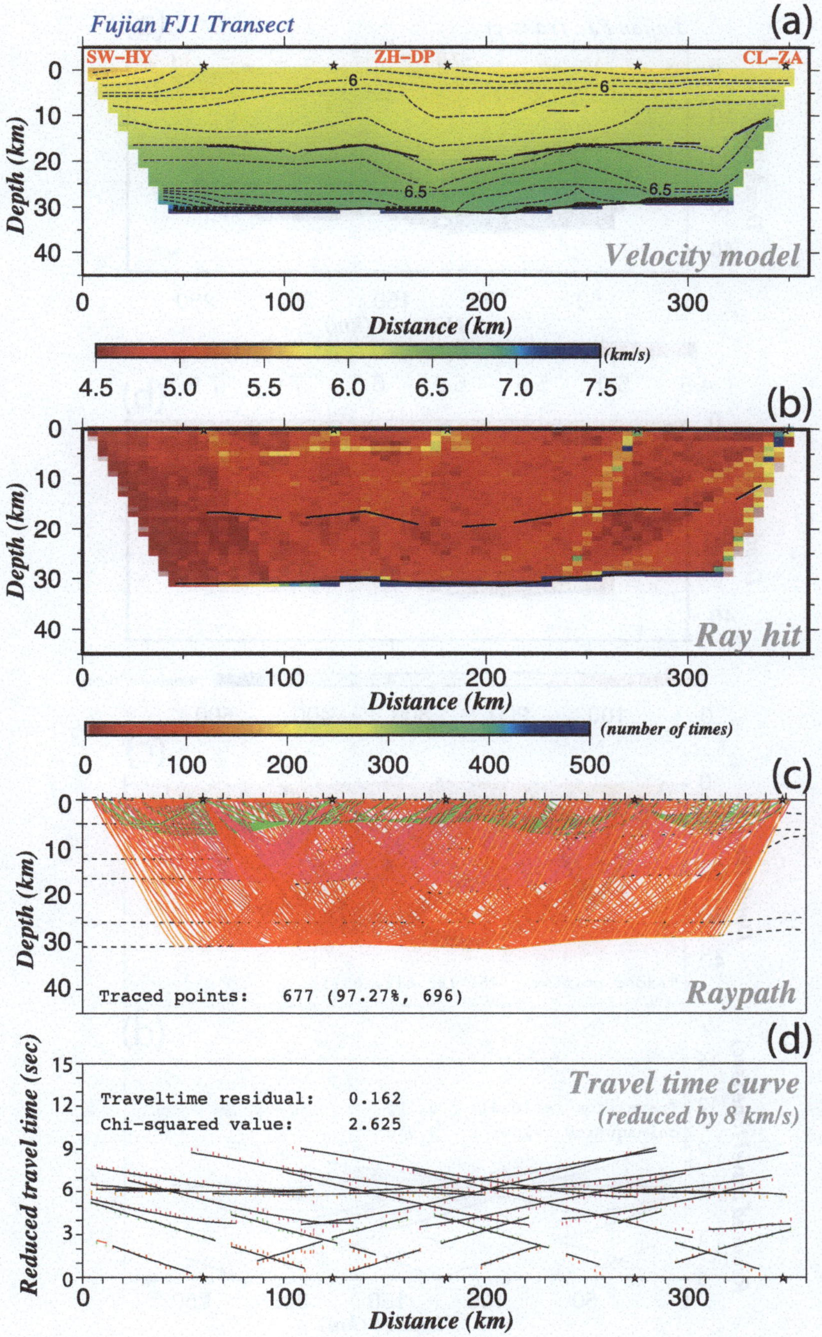

Fig. 8.3 The P-wave velocity modeling of the FJ1 transect: **a** velocity model, **b** ray hits, **c** ray paths, and **d** travel-time residual and chi-squared values

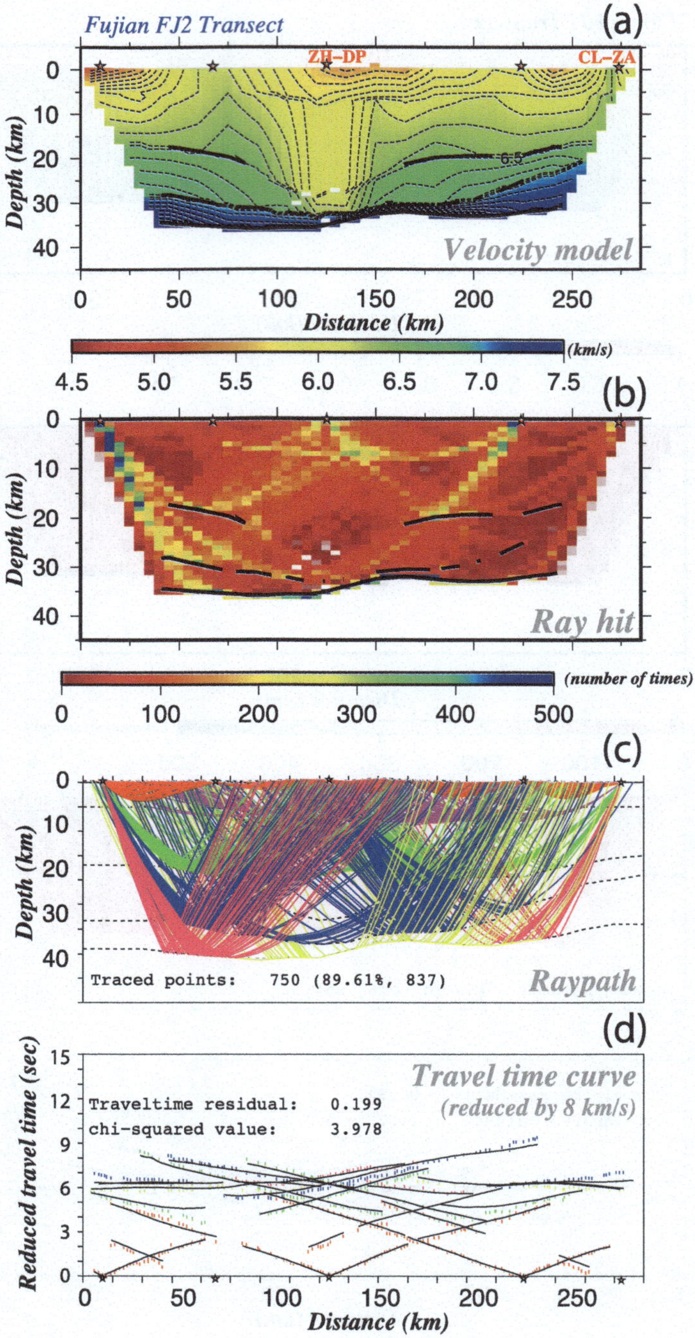

Fig. 8.4 The P-wave velocity modeling of the FJ2 transect: **a** velocity model, **b** ray hits, **c** ray paths, and **d** travel-time residual and chi-squared values

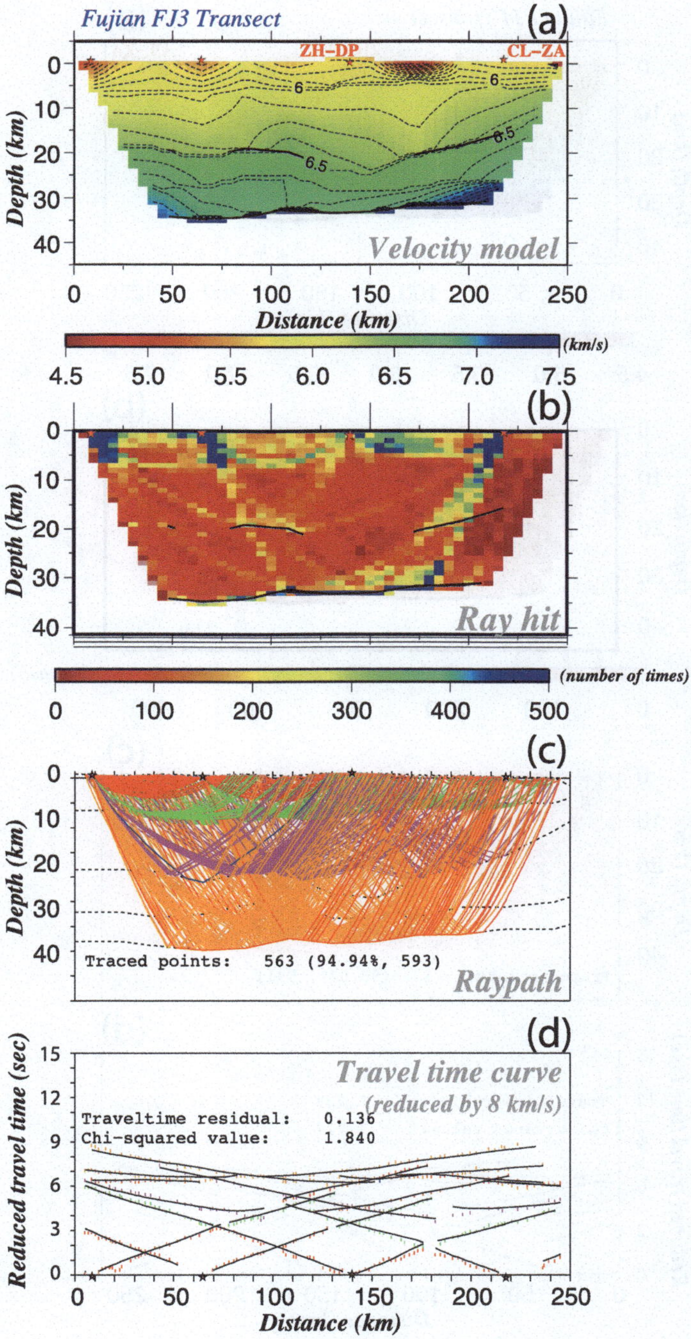

Fig. 8.5 The P-wave velocity modeling of the FJ3 transect: **a** velocity model, **b** ray hits, **c** ray paths, and **d** travel-time residual and chi-squared values

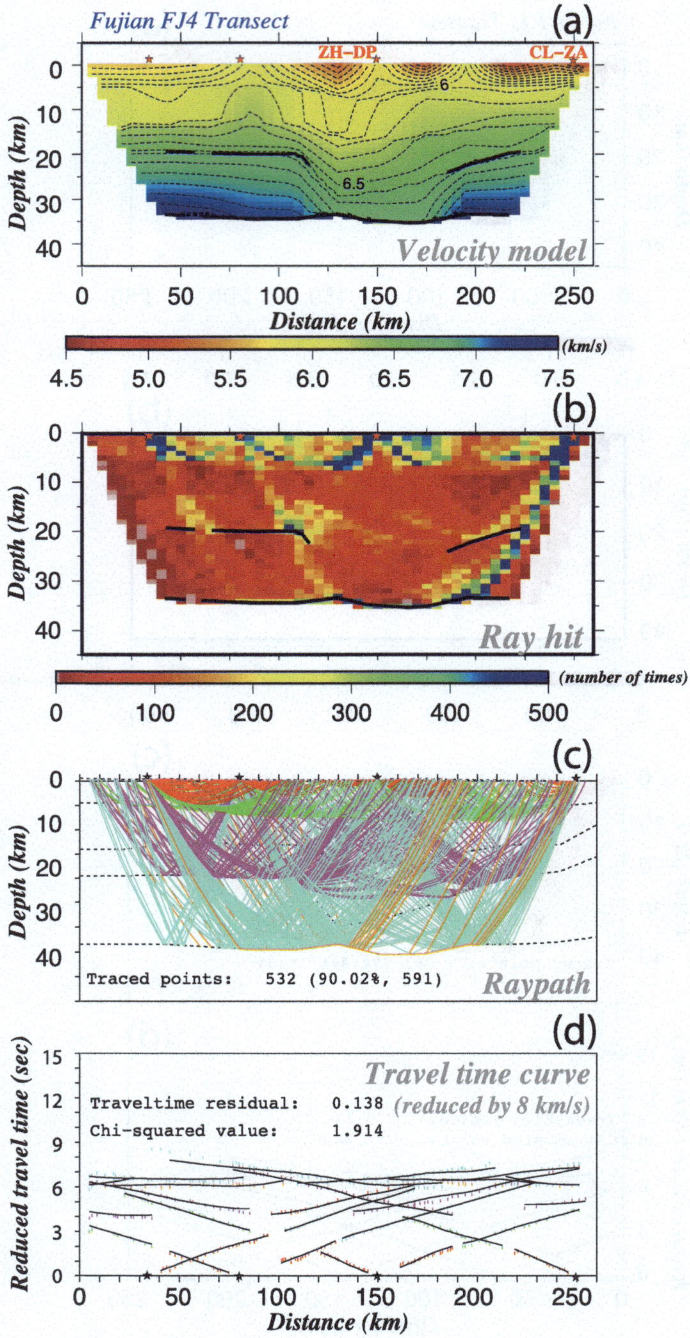

Fig. 8.6 The P-wave velocity modeling of the FJ2 transect, **a** velocity model, **b** ray hits, **c** ray paths, and **d** travel-time residual and chi-squared values

8.3 Magnetic and Gravity Anomalies

Seismic velocity profiles revealed a relatively low-Vp velocity pattern along the Zhenghe-Dapu fault zone (Fig. 8.7). To further understand this special feature, magnetic and gravity anomaly data were used to provide other geophysical constraints beneath the study area and to reflect the physical properties of the underlying rocks. As the magnetized field had a 90° inclination, to map the geomagnetic characteristics of the Fujian province, the magnetic data from Doo et al. [4] are applied to the reduction to the pole (RTP) technique. The magnetization direction is presumed to be the same as in the present-day geomagnetic field (inclination is set to 38.98° and the declination is set to −3.86° based on the IGRF11 model: http://www.ngdc.noaa.gov/IAGA/vmod/igrf.html) in the study area. To focus on the deeper (and/or large-scale) structural characteristics, we use a low-pass filter (G.M.T. software) to remove the short-wavelength signals (wavelengths smaller than 30 km); the result is shown in Fig. 8.8a. The magnetic high trend is roughly parallel to the Mainland China shoreline, which is also identical to the major structural direction in southeast China. Remarkably, a negative magnetic anomaly zone was located in the central part of Fujian province, and the trend is nearly NE–SW in direction. Moreover, the location of this negative magnetic anomaly zone is consistent with the Zhenghe-Dapu fault zone as well as the low-Vp velocity zone (Fig. 8.7).

To expose the density feature in the deeper part of the study area, a low-pass filter is applied to process the Bouguer gravity anomalies [1]. The filtering parameters used here are the same as those used to process magnetic data, and the result is shown in Fig. 8.8b. As a result, the NE–SW-trending low-Bouguer gravity anomaly zone is very significant, indicating the existence of low-density material or crustal thickening. According to Birch's law, there exists an essentially linear relationship between seismic velocity and density. We can see that the locations of low-Vp velocity (Fig. 8.7) and low-Bouguer gravity anomalies (Fig. 8.8b) show a good match in the Fujian province. This fault zone is suggested to be a rift basin filled with a chaotic sedimentation. Sediments generally reveal low susceptibility, low density, and relative weakness compared to both sides of the rifted basin. The compressional force may have forced these low-susceptibility and low-density materials to deeper depths in the Fujian province, as shown in the seismic profiles (Fig. 8.7).

8.4 Discussion and Conclusion

Because of the rapid growth in isotopic and geochronological data sets, the tectonic history of the SCC has been constructed in much more detail (Fig. 8.9). During the Mesoproterozoic era, the Yangtze and Cathaysia blocks belonged to different terranes, and their own interiors contained micro-plates and/or micro-continents. Around 1000–860 Ma, the Neoproterozoic collisional orogeny assembled these two blocks to form the ancient SCC as part of the Rodinia supercontinent [10–12, 15,

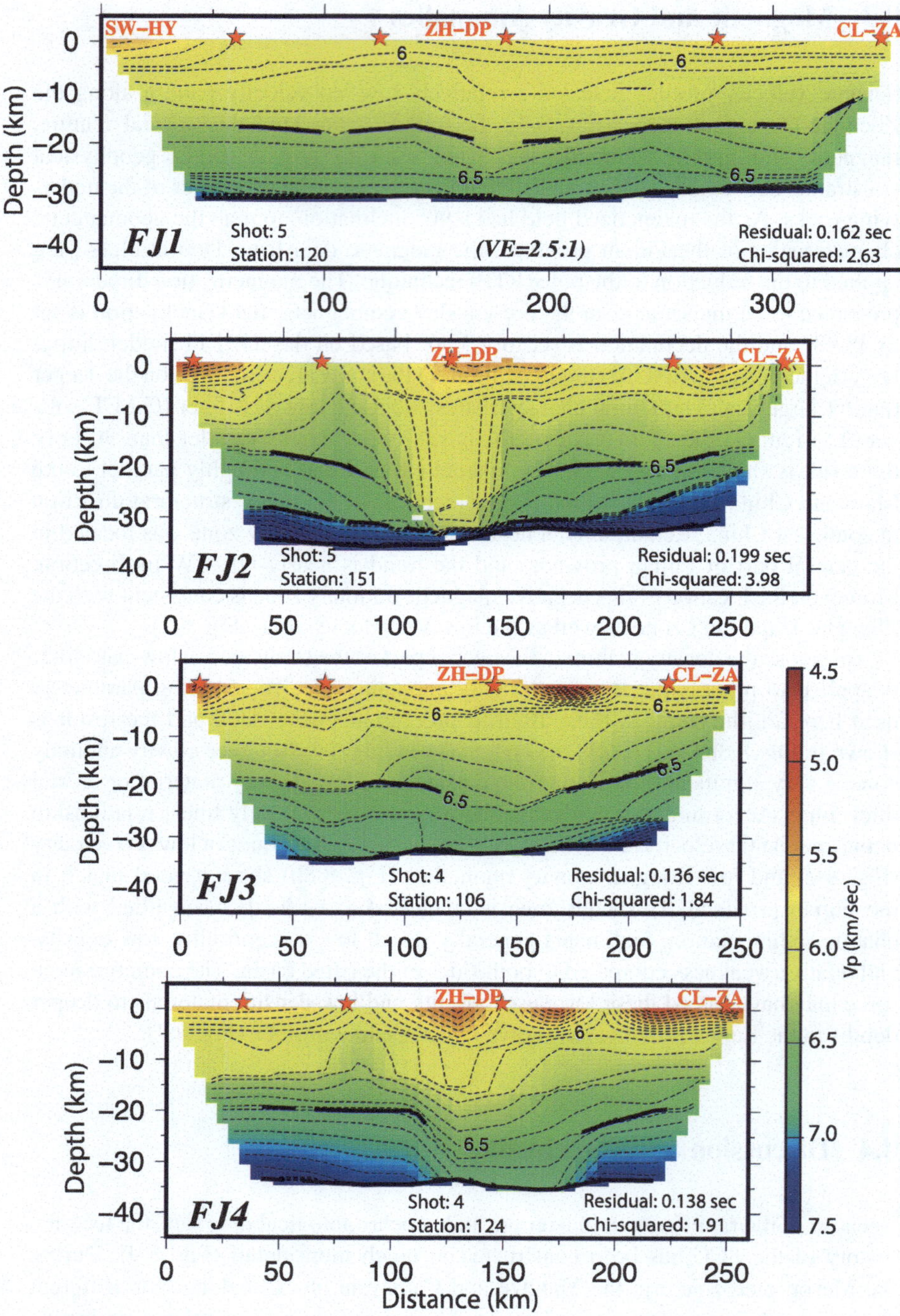

Fig. 8.7 Reflection/refraction P-wave velocity models. Thick black line: reflector

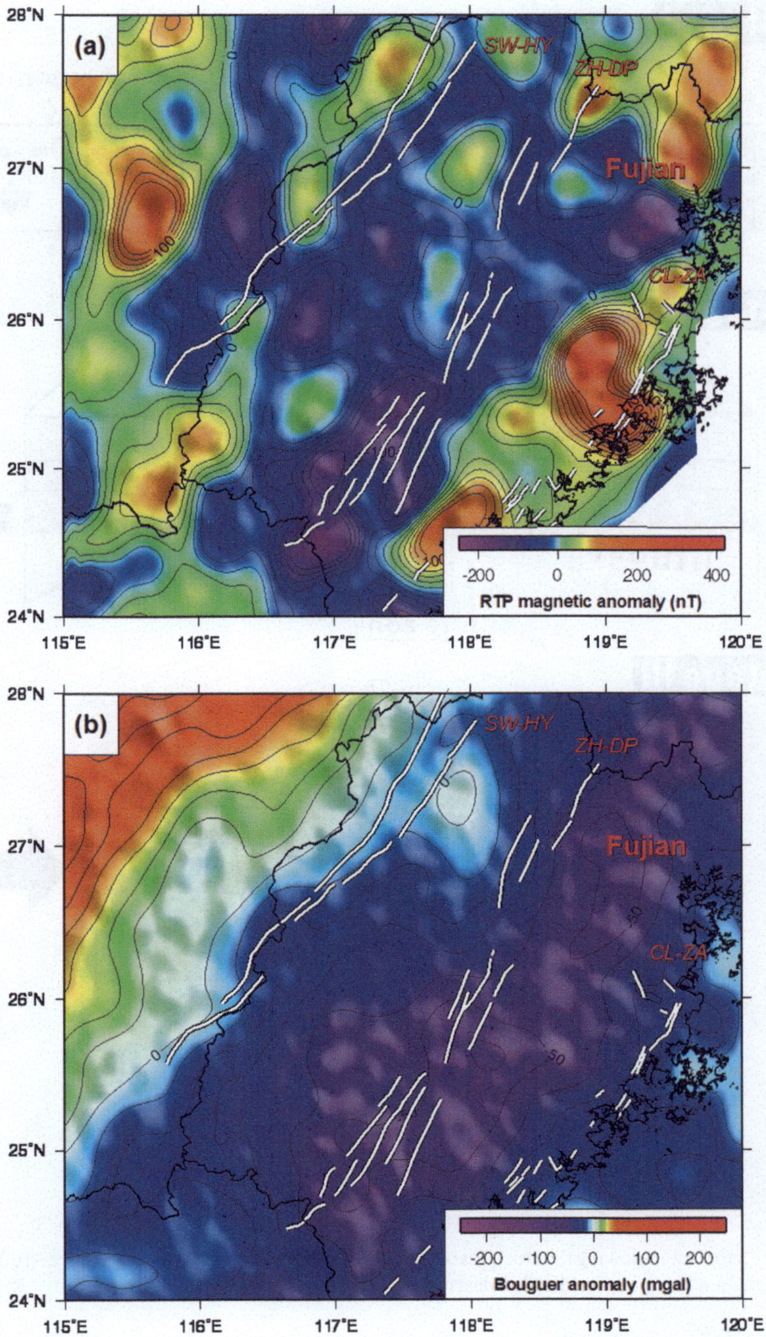

Fig. 8.8 **a** Magnetic anomalies from Doo et al. [4]. **b** Bouguer anomalies from Bonvalot et al. [1]. White line: fault zone. See Fig. 8.2 for the full names of the faults

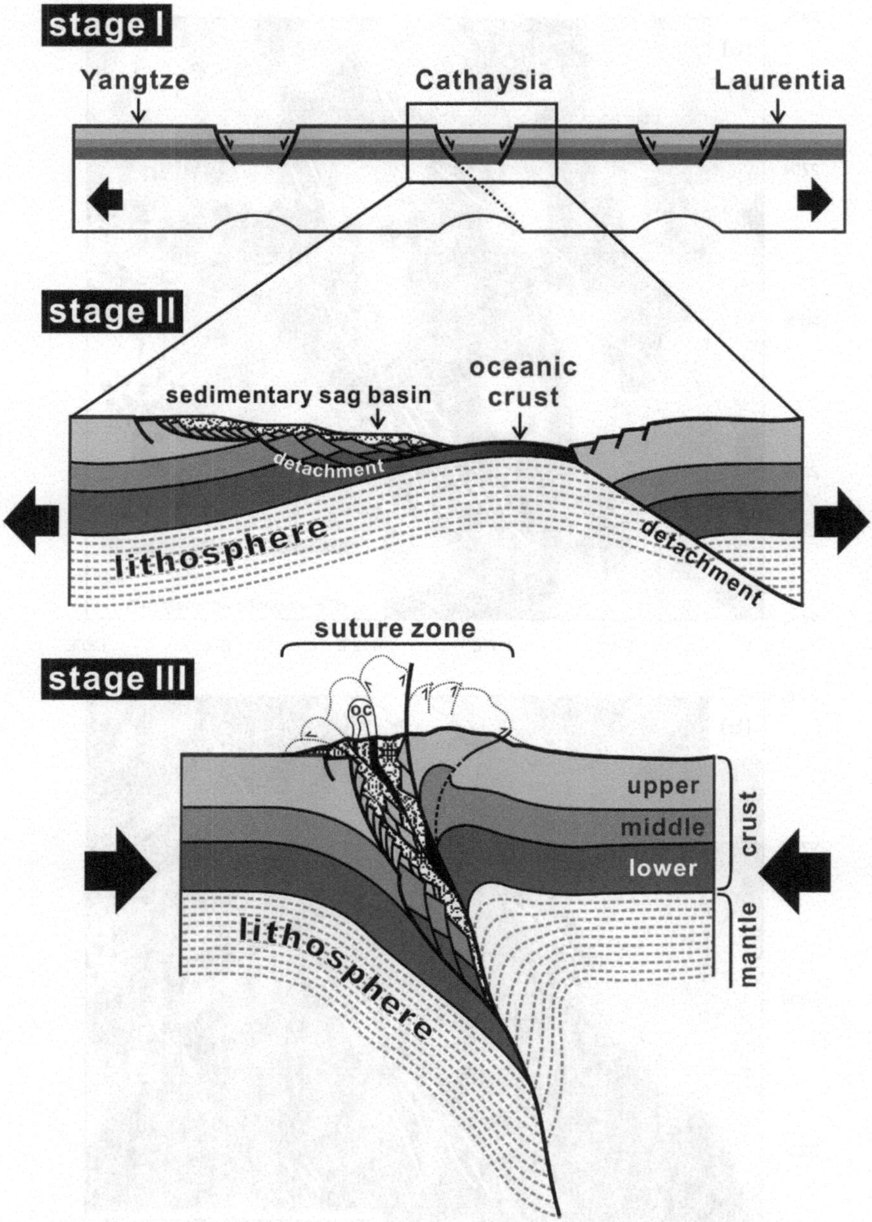

Fig. 8.9 Schematic showing the crustal evolution of the Zhenghe-Dapu suture zone since ~1000 Ma. See the main text for a more detailed description. OC: oceanic crust

20, 22]. Previous studies have suggested that the Cathaysia block, which has a more complicated tectonic history than the Yangtze block, comprises eastern and western Cathaysia blocks that represented separate micro-continents due to the breakup of the Rodinia supercontinent ca. 860–800 Ma [14] (Stage I of Fig. 8.9), and even before the Indosinian (ca. 240 Ma) or at least the Caledonian (ca. 450 Ma) events, these two blocks were kept separated [17, 20] (Stage II of Fig. 8.9). Until the Yanshanian event (ca. 160 Ma), these two terranes joined along the Zhenghe-Dapu fault zone [3, 17] (Stage III of Fig. 8.9). This fault zone is suggested to be a rift basin filled with chaotic sedimentation along with ultramafic blocks, pillow basalts, gabbros and dykes, acidic volcanic rocks, or, more generally, the bimodal igneous rocks such as those exposed in the Zhenghe area. The age of initial sedimentation within the Zhenghe-Dapu rift basin was estimated to be approximately 860 Ma, and all of the rocks deposited in this rift basin were ductility deformed and metamorphosed during the early Paleozoic intra-continental orogeny of South China [5, 14].

Although the geological evidence for this rift basin is distributed only on the surface near both sides of the fault zone, owing to its long tectonic history, as the dramatic deformation from a rift system to a continental collision in this suture zone shows, the crustal structure from geophysical data sets reveals the significant imprints of this structure and advances our knowledge of the Cathaysia plate reconstruction.

Acknowledgments The authors thank the Earthquake Administration of Fujian Province and the Geophysical Exploration Center of the China Earthquake Administration for their laborious efforts to collect field data. H. K-C was supported by the National Science Council of Taiwan (Grant No. NSC101-2116-M-008-023-MY3). X. Jin and H. T. Cai were supported by the National Natural Science Foundation of China (Grant No. 41474071). Initial discussion with K.-L. Wang of the Institute of Earth Science of Academia Sinica is highly appreciated. The authors would also like to thank Uni-edit (www.uni-edit.net) for editing and proofreading this manuscript.

References

1. Bonvalot S, Balmino G, Briais A, Kuhn M, Peyrefitte A, Vales N, Biancale R, Gabalda G, Reinquin F, Sarrailh M (2012) World gravity map. Commission for the Geological Map of the World. Eds. BGI-CGMW-CNES-IRD, Paris
2. Cai HT, Kuo-Chen H, Jin X, Wang CY, Huang BS, Yen HY (2015) A three dimensional Vp, Vs, and Vp/Vs crustal structures in Fujian, Southeast China, from active- and passive-sources ecperiment. J Asian Earth Sci. https://doi.org/10.1016/j.jseaes.2015.06.014
3. Chen J, Jahn B-M (1998) Crustal evolution of southeastern China: Nd and Sr isotopic evidence. Tectonophysics 284:101–133
4. Doo W-B, Hsu S-K, Armada L (2015) New magnetic anomaly map of the East Asia and some preliminary tectonic interpretations. Terr. Atmos. Ocean. Sci. 26:73–81
5. Faure M, Shu LS, Wang B, Charvet J, Choulet F, Monié P (2009) Intracontinental subduction: a possible mechanism for the Early Palaeozoic Orogen of SE China. Terre Nova 21:360–368
6. He C, Dong S, Santosh M, Chen X (2013) Seismic evidencd for a geosuture between the Yangtze and Cathaysia blocks, south China. Sci Rep 3. https://doi.org/10.1038/srep02200
7. Hole JA (1992) Nonlinear high-resolution three-dimensional seismic travel time tomography. J Geophys Res 97:6553–6562

8. Huang Z, Wang L, Zhao D, Xu M, Mi N, Yu D, Li H, Li C (2010) Upper mantle structure and dynamics beneath Southeast China. Phys Earth Planet Inter 182:161–169
9. Legendre CP, Deschamps F, Zhao L, Lebedev S, Chen Q-F (2014) Anisotropic Rayleigh wave phase velocity maps of eastern China. J. Geophys. Res. Solid Earth 119:4802–4820. https://doi.org/10.1002/2013JB010781
10. Li X, McCulloch MT (1996) Secular variation in the Nd isotopic composition of Neoproterozoic sediments from the southern margin of the Yangtze Block: Evidence for a Proterozoic continental collision in southeast China. Precambrian Res. 76:67–76. https://doi.org/10.1016/0301-9268(95)00024-0
11. Li X, Li ZX, Zhou HW, Liu Y, Kinny PD (2002) U-Pb zircon geochronology, geochemistry and Nd isotopic study of Neoproterozoic bimodal volcanic rocks in the Kangdian Rift of south China: Implications for the initial rifting of Rodinia. Precambrian Res. 113:135–154. https://doi.org/10.1016/S0301-9268(01)00207-8
12. Li X, Li WX, Li ZX, Liu Y (2008) 850–790 Ma bimodal volcanic and intrusive rocks in northern Zhejiang, south China: A major episode of continental rift magmatism during the breakup of Rodinia. Precambrian Res. 102(1–2):341–357
13. Li X (2013) Subdivision and characteristic of tectonic units in Fujian Province. Global Geology 32(3):549–557 (in Chinese with English abstract)
14. Shu LS, Faure M, Yu JH, Jahn BM (2011) Geochronological and geochemical features of the Cathaysia block (South China): new evidence for the Neoproterozoic breakup of Rodinia. Precambrian Res. 187:263–276
15. Wang, Y., Zhang, F., Fan, W., Zhang, G., Chen, S., Cawood, P.A., Zhang, A., 2010. Tectonic setting of the South China Block in the early Paleozoic: Resolving intracontinental and ocean closure models from detrital zircon U-Pb geochronology. Tectonics 29, TC6020, https://doi.org/10.1029/2010tc002750
16. Wong J, Sun M, Xing G, Li X-H, Zhao G, Wong K, Wu F (2011) Zircon U-Pb and Hf isotopic study of Mesozoic felsic rocks from eastern Zhejiang, South China: Geochemical constrast between the Yangtze and Cathaysia blocks. Gondwana Res 19:244–259
17. Xu XS, O'Reilly SY, Griffin WL, Wang XL, Pearson NJ, He ZY (2007) The crust of Cathaysia: Age, assembly and reworking of two terranes. Precambrian. Res. 158:51–78
18. Zelt CA, Smith RB (1992) Seismic traveltime inversion for a 2D crustal velocity structure. Geophys J Int 108:16–34
19. Zelt CA, Forsyth DA (1994) Modeling wide-angle seismic data for crustal structure:Southeastern Grenville province. J Geophys Res 89:11687–11704
20. Zhang G, Guo A, Wang Y, Li A, Dong Y, Liu S, He D, Cheng S, Lu R, Yao A (2013) Tectonics of South China continent and its implications. Sci China Earth Sci 56, 1804–1828. https://doi.org/10.1007/s11430-013-4679-1
21. Zhao B, Bai Z, Xu T, Zhang Z, Badal J (2013) Lithological model of the south China crust based on integrated geophysical data. J Geophys Eng 10. https://doi.org/10.1088/1742-2132/10/2/025005
22. Zhou XM (2003) My thinking about granite geneses of south China. Geol J China Univ 9(4):556–565 (in Chinese with English abstract)

The manufacturer's authorised representative in the EU is Springer Nature Customer Service Centre GmbH, Europaplatz 3, 69115 Heidelberg, Germany. If you have any concerns regarding our products, please contact ProductSafety@springernature.com

Printed and bound by CPI Group (UK) Ltd, Croydon, CR0 4YY
30/09/2026
02225574-0008